BAMBOOS OF BHUTAN

AN ILLUSTRATED GUIDE

Chris Stapleton

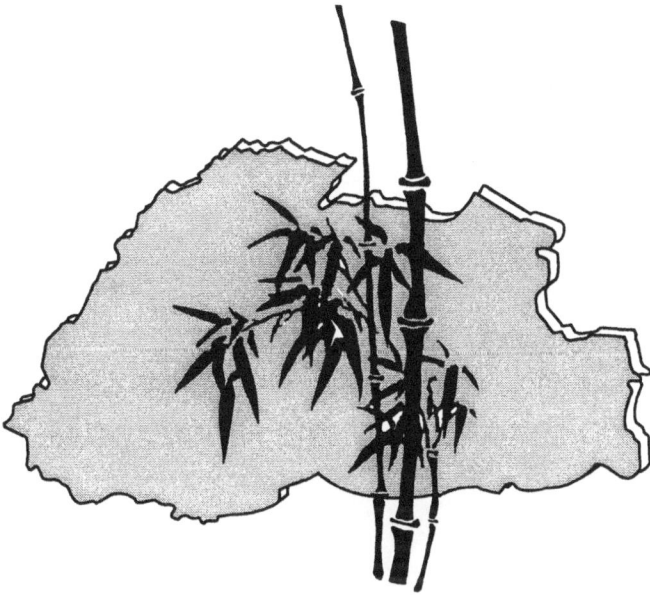

Illustrations of the genera and species, with notes on identification, distribution, utilisation, and propagation

BAMBOOS OF BHUTAN:
AN ILLUSTRATED GUIDE

Chris Stapleton

Forestry Department, University of Aberdeen
Royal Botanic Garden Edinburgh
Royal Botanic Gardens Kew

in association with

Forest Research Division
Forest Department
Royal Government of Bhutan
Thimphu

Royal Botanic Gardens, Kew, on behalf of

The Overseas Development Administration, London

Forestry Research Programme, University of Oxford

Published by The Royal Botanic Gardens, Kew

for   The Overseas Development Administration of the British Government
      Forestry Research Programme
      University of Oxford,
      Halifax House,
      6 South Parks Road,
      Oxford OX1 3UB

First published 1994

Design, illustrations, and layout by the author. Cover by Media Resources, RBG Kew.

Research for this guide and its production were funded by the Overseas Development Administration, under research grants R4195 and R4849. Field work was implemented by the Forestry Department of Aberdeen University in conjunction with the Forest Department of the Royal Government of Bhutan. Illustrations and camera-ready copy were produced at the Royal Botanic Garden Edinburgh. Final editing and production were supported by the Anglo-Hong Kong Trust.

ISBN 0 947643 67 2

Printed in Great Britain by Whitstable Litho Ltd.

CONTENTS

INTRODUCTION

BAMBOOS are widely distributed across the country, but they are most common in higher rainfall areas. They can be found from the subtropical southern borders to the tree-line, where they form extensive areas of pasture for yak and sheep. Tropical species from Malaysia and Burma extend across Assam into southern Bhutan. Temperate and sub-alpine genera that are more common in Tibet can be found at altitudes of up to 4,000m in the North.

The poles of several species are important minor forest products, and they are systematically harvested on an annual basis. Many other species are widely planted on private land. While traditional uses continue to use large quantities of bamboo products, new uses are now also being developed, and substantial new export markets remain largely unexplored. Demand for bamboo poles for building construction in India is very large, and increasing shortages of timber make bamboo more valuable. Low-cost housing consumes large quantities of woven bamboo panelling, and construction of modern buildings requires more bamboo for scaffolding. New uses include bamboo reinforcement for concrete, lamination into boards and utensils, and canning of the shoots of edible species.

There are very many uses for bamboos within Bhutan. Roof mats and fencing are the main uses at higher altitudes. In subtropical areas where the larger species will grow bamboo is treated as a multipurpose raw material from which almost anything can be made. Most bamboo products however,

are not very durable, and they are attacked by fungi and insects faster than wooden products, so that they need to be replaced on a regular basis. The flexibility of many bamboo species is an important characteristic, allowing the weaving of thin strips into all manner of baskets and trays. These are used for collecting, sorting, storing, and then transporting most agricultural products.

The role of bamboos in soil conservation can be considerable in the Himalayas. Because of their dense surface roots, large subtropical bamboos and even small temperate species protect against sheet and gully erosion more effectively than most trees. Bamboos along roadsides can reduce the surface erosion and the small landslides that fill the drains or block the roads. The rhizome system of larger bamboos can form an effective buttress, holding up terraces and road banks. In combination with selected trees that can root to a greater depth, bamboos provide a low cost means of slope stabilisation that can also provide useful products.

Against the usefulness of most bamboos, there are a small number of species that can restrict regeneration of timber trees. When bamboos grow in clumps, tree seedlings can usually grow in the gaps between the clumps, but spreading genera such as *Yushania* cover the ground more completely. Under natural conditions mature trees prevent enough light reaching the ground for bamboo to grow too vigorously, and tree seedlings can grow through the bamboo. However, if large sections of the tree canopy are removed completely by clear-felling, larger *Yushania* species become very dense and tree seedlings can no longer grow through them.

The identification of bamboos has been rather neglected in the past. Bamboos are giant grasses, but they differ from the smaller grasses in several ways. Unlike all other grasses, they have woody culms, well-developed branching, specialised culm sheaths, and leaf bases narrowed into thin petioles. They also have cyclical flowering. For a long time taxonomists thought that flowers were the most important parts of bamboo plants for identification purposes, but it is now accepted that their vegetative characteristics are equally important. As some species may wait up to 150 years before flowering, this makes it much easier to identify the different genera and species.

There has recently been a period of great confusion over the genera of Himalayan bamboos. The smaller species were originally all placed in the genus *Arundinaria*. Over the last century and during the 1980's in particular, many new genera were described in Japan and China. These genera were not always clearly defined, and so were often not recognised in other countries. Now that much more information on Sino-Himalayan bamboos is available to the scientific world, many of these new genera are becoming accepted. However certain new genera, such as *Sinarundinaria* and *Butania,* have been shown to be the same as genera which had already been named. A more stable system of genera is now recognised, and is it hoped that it will not be necessary to make changes to generic names again.

The species of bamboos in Bhutan are still not very well known, and the number of collections made is still very small in many species. Several other species probably remain undiscovered in less accessible parts of the country.

Hopefully most of the more common species from central Bhutan have been included in this guide. More accurate information is still needed on the distribution, uses, and local names of all the species. This guide should form a framework within which a much more comprehensive account can be written by Bhutanese taxonomists.

Species are described in this guide to allow positive identification in the field from vegetative material alone. It is aimed at forestry and agricultural personnel rather than just specialised taxonomists, therefore the terminology is kept as simple as possible. Accurate identification of species does require a little detailed knowledge of the different parts of bamboo plants. Therefore the most important parts are briefly described. Important features of each genus and each species are illustrated. The genera are separated using a key, and a full glossary is given. Individual species are described, showing how to distinguish them from closely related species, and giving some basic information on distribution, uses, and the most appropriate propagation techniques. Species from neighbouring countries are sometimes included to help to distinguish Bhutanese species.

No account of bamboos would be complete without some reference to their peculiar flowering behaviour. It is now known that different species have very different flowering habits. Lengths of flowering cycle vary greatly between varieties and species, and not all species die after flowering. It is not yet possible to list the flowering habits of each species, but hopefully this guide will allow better records to be kept of the flowering patterns of different species in the future.

IDENTIFYING BAMBOOS

To identify bamboo species the most important parts of the plant are the culm sheaths. These are protective sheaths around the stems (the stem is called a culm in all grasses), see fig. 1. The sheaths below the leaves (leaf sheaths) are also important, see fig. 2). At the top of these sheaths there is a projecting tongue in the centre called the ligule, and ears on each side called auricles. The shape and size of the

Culm sheaths at the culm base are different from those higher up. They are broader and have shorter blades. To standardise descriptions, culm sheaths at eye-level on the large bamboos are taken. These are approximately ¼ of the way up the culm. Smaller bamboos are treated in the same manner, culm sheaths from ¼ of the way up the culm from the base being described.

New culm sheaths show the features of the species best. Older sheaths often have parts that are missing or have rotted away, especially in hotter areas,

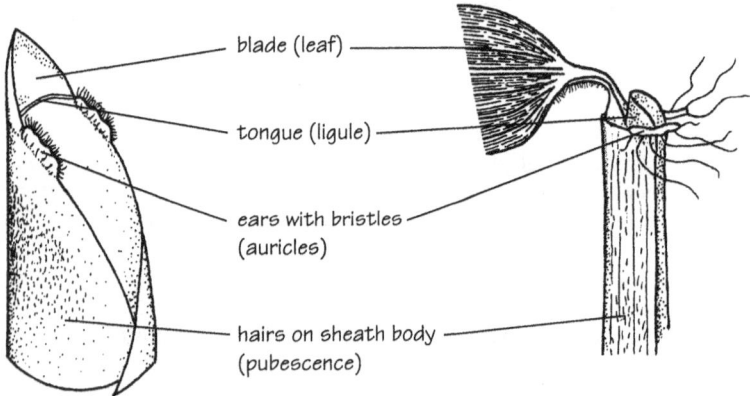

blade (leaf)

tongue (ligule)

ears with bristles
(auricles)

hairs on sheath body
(pubescence)

fig. 1 - culm sheath   fig. 2 - leaf sheath

auricles, and whether there are stiff bristles on their edges are all important. The shape, length and the type of edge on the ligule are also important.

The blade of the culm sheath is a modified leaf. Its shape, whether it has hairs on the back or around the base, whether it is erect or bent backwards (reflexed), and whether it falls off early (deciduous), or will remain attached (persistent), are also all important.

and for this reason bamboos are easiest to identify in the late summer and autumn. In winter and spring care must be taken to find undamaged sheaths. In the same way, leaf sheaths are damaged by strong wind and rain so that auricles, bristles, and hairs are blown off after a few months. Small new leaves can be found at most times of year except in winter. Some of these should be collected as well as larger, older leaves. The drawings in this guide show fairly new parts. These are typical of those which would be found in October.

The surface of the culm is also important. Young culms have a coating of wax, which can be either thick and furry, or thin, and either light or dark in colour, and it may rub off quickly to leave the culms shiny, or it may persist so that the culms stay matt and dull. The joints of the culm (nodes) may be raised or level, with rings of different colours, and they may bear small aerial roots or thorns. The surface of the culm may be rough with tiny sharp points, or smooth, or it may be covered in small vertical ridges.

Branching is a useful and important characteristic in bamboos, especially in the separation of genera. The number of branches in the first year of growth is important, as well as the eventual number of branches which older culms develop. Whether these branches are all the same size should be noted, or whether the central branch is much larger than all the others, in which case it may develop aerial roots on its base.

Rhizomes are difficult to examine as they usually remain under the ground. The type of rhizome will determine whether the bamboo grows in a clump (clump-forming), or spreads widely (running). Running rhizomes may have roots at all nodes, or they may have roots only on the short internodes near the bases of the culms, and they may be solid or hollow.

Flowers of bamboos are occasionally found. There are two different kinds of bamboo inflorescence. One type will keep on branching to give dense clusters or rounded balls of flowers, which are well-developed in genera *Dendrocalamus* and *Bambusa,* and in *Cephalostachyum.*

The other type produces flowers in large panicles, similar to those of an ordinary grass. The panicles and flowers of *Thamnocalamus* remain partially hidden by sheaths, while the sheaths fall quickly from the panicles of other genera such as *Drepanostachyum* and *Borinda.*

The colour of the flowers can allow quick identification of the large bamboos if the flowers are young, but they will all fade to a brown or straw colour. *Dendrocalamus hamiltonii* var. *hamiltonii* has purple flowers with distinctive red anthers, while var. *edulis* has yellow flowers. *Dendrocalamus hookeri* has olive-green to brown flowers. *Bambusa tulda* and *B. nutans* have green flowers, while *Bambusa balcooa* has green flowers with purple tips. *Bambusa clavata* has large purple flowers with yellow anthers. *Dendrocalamus giganteus* has very long pendulous sprays of flowers.

To allow accurate identification of bamboos in the herbarium, a collection of leaves and culm sheaths is usually adequate if they are in good condition and well protected. If a proper plant press is not available, the leaves can be packed inside a rolled culm sheath, and a series of culm sheaths can be rolled together and tied firmly. The outer sheaths will protect the delicate parts of those at the centre, such as the auricles and blades. For the small bamboos a section of the culm is very useful, including a node with its branches cut back to 5cm. For spreading bamboos a short section of the rhizome is also required. This can often be found on an over-hanging bank or road-cutting. If flowers are collected, leaves and culm sheaths should always be searched for and included, even if old. Collections should never be put in plastic bags as they will rapidly go mouldy.

PROPAGATION METHODS

HIMALAYAN bamboos can be propagated either from seed, or by vegetative methods. Seed should always be used when a suitable species or variety is flowering, as long as there is a nursery in which the seedlings can be grown. This guarantees the maximum period of vegetative growth before flowering begins on the planted material. Various techniques of vegetative propagation can also be used when seed is not available. The traditional propagation technique, which is essentially clump division, and involves digging out a section of rhizome, has to be used when nursery facilities are not available. Other kinds of cuttings can be used when there is a reliable nursery nearby. Different forms of propagation or type of cutting are appropriate for different genera and species. Plants raised in nurseries either from seed or from cuttings can be produced in large quantities, but being smaller than the traditional planting material, they require better protection from grazing animals.

RAISING SEEDLINGS

Collection of seed from flowering clumps is best organised by local private seed collectors. It requires local knowledge of where and when bamboos are flowering, and a rapid response to obtain good quality seed before it is destroyed by insects, rain or fire. Seed of *Bambusa* and *Dendrocalamus* species may be produced within three weeks of the start of flowering. Unlike agricultural grasses, which have been bred to retain their seed, the seed of bamboos will often fall to the ground as soon as it is mature. Collection of good seed involves collection of the seed as it falls by placing sheets or tarpaulins underneath the flowering clumps, or collection of the fallen seed from the litter and vegetation on the ground. Collection of the flowering branches usually results in loss of most of the viable seed as it is so easily dislodged, although some temperate bamboos retain their seed in the flowers for longer than the subtropical bamboos.

The seed should be dried in the sun and cleaned. The chaff can be separated from the seed by gentle rubbing and winnowing. Insects may destroy the seed completely within a few months if they are not eliminated. The principal pest is *Sitotroga cerealella*, a small light brown moth with tiny larvae that burrow into the seed. They eat the seed contents, leaving white papery remnants of their cocoons, and they can complete their life cycle in 5 weeks. Treatment with insecticide powder or placing the dried seed in a freezer for 3 days is necessary to control this pest.

Storage of bamboo seed is very difficult, even after elimination of insect pests. It can be dried, but even when the moisture content is reduced to the ideal level of around 8-10%, and the temperature is maintained at 5°c, the germination rates will still fall to 25% or less in the first year. This means that some seed may be stored to be sown in a second season, but most of it should really be sown before the first monsoon. Longer periods of storage are possible if dried seed could be stored at a constant -18°c, but repeated thawing and freezing is likely to kill the seed.

Seed of large subtropical bamboos such as *Dendrocalamus hamiltonii* has no dormancy and fresh seed will normally germinate within two weeks if conditions of temperature and humidity are suitable. Seed of the smaller subtropical and temperate bamboos may have substantial dormancy, and it might germinate more quickly after a period of cold pre-treatment, such as stratification or refrigeration at 5°c. *Himalayacalamus hookerianus* seed stored and sown at 20-25°c. germinated very slowly over a nine month period, the first shoot appearing on the 35th day after sowing.

Seed can be sown into seedbeds or it can be sown directly into containers. Seedlings of *Dendrocalamus strictus* can be grown without shading. The seedlings of most of the Himalayan bamboos require good shading and frequent watering, and they may be difficult to transplant without loss.

As it is not always possible to collect the seed in time, it may sometimes be necessary to collect the germinated seedlings from under flowering clumps in the monsoon after flowering. Very dense regeneration is sometimes seen, if the pressure from grazing animals is low and there are no fires to destroy the seed. These natural seedlings can be transplanted directly into containers in shaded beds, but they need to be kept moist during transportation, and they will lose their leaves or die if they are not transplanted promptly.

TRADITIONAL PROPAGATION

Only one technique of propagation has been used on any scale in the Himalayas in the past. An accessible culm from near the edge of the clump is removed by digging around its rhizome, cutting the rhizome where it branches from its mother. The length of the culm is reduced by cutting it 1-2m above the ground. This propagation technique is undertaken at the start of the monsoon. If undertaken too early the roots will dry out and the rhizome will die. If left too late buds on the rhizome will have already developed into fragile new shoots, which will die during the transplanting process. If the operation is successful, the culm will grow new

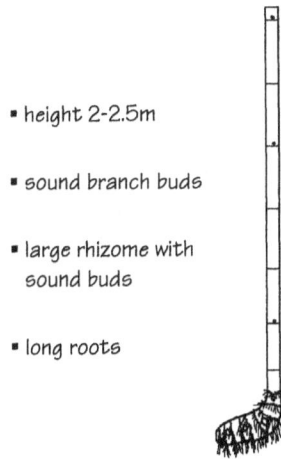

- height 2-2.5m

- sound branch buds

- large rhizome with sound buds

- long roots

fig. 3 -traditional planting material

branches and leaves at the top in the first year, and in the second year a new shoot will emerge from the rhizome, hopefully reaching a height of several metres. Small bamboo species should be planted using several culms and rhizomes still joined together.

This is a robust planting method, which can establish bamboo clumps in areas where there is substantial grazing of livestock. The drawbacks are the high labour costs, and the shortage of planting material available, which both limit the scale upon which it can be implemented. However, this is probably the only technique which will be successful in planting areas where grazing cannot be prevented.

It is possible to improve the success rates and the speed of establishment achieved using the traditional technique. Selection of older culms with more reserves in their rhizomes will provide better tolerance to drought and grazing. However, older rhizomes are more difficult to extract. Use of a longer culm section of 2-2.5m reduces the browsing of the new growth at the top of the culm, but this makes transport more difficult. Support of the culm with two poles in a tripod arrangement will prevent animals from pushing it over to reach the leaves. Watering of the roots during extraction, transport, and during periods of drought will greatly improve growth rates. Protection of new shoots and foliage with branches from thorny bushes may reduce the damage caused by animals.

CULM CUTTINGS

Many species of *Dendrocalamus* and *Bambusa* produce aerial roots from the bases of larger branches. The rooting of branches can be used to raise new plants without the extraction of rhizomes. Various propagation techniques based upon the rooting of branches are used in areas of the tropics with heavy spring rainfall. They are not commonly used in monsoonal areas with a spring drought. This is because branches grow in the spring, and without moisture they cannot root effectively. Where nursery facilities are now available to provide abundant artificial watering from spring until the monsoon, culm cuttings of many species can be highly successful. As well as frequent and regular watering,

cuttings also require good shading and protection from grazing animals.

Branches on their own rarely have sufficient reserves to sustain strong new shoots until they root. Sections of the culm with branch bases attached are more successful. The principal limiting factor in the growth of new shoots is usually water availability. Single-node culm sections planted horizontally with both ends buried expose a large area of vascular culm tissue to the wet soil. This maximises the entry of water first into the culm section, and then into each branch and its buds.

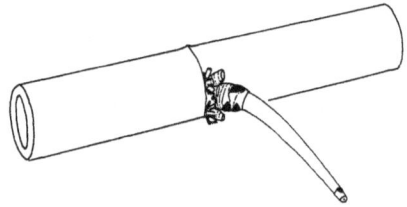

fig. 4 - single-node culm cutting

Cuttings are taken just before growth of new branch shoots and leaves in the spring, normally from mid-March to mid-April. Large 2-year-old culms with strong branches should be chosen. These culms would often be harvested during the previous winter, so it is advisable to buy and mark them earlier. The buds at the base of the central branches must not be damaged. The culms are felled and the branches are trimmed back. The central branch is cut at a length of about 20cm, beyond the first long internode, while the smaller branches are cut right back to the culm. The culm is then cut into single-node sections, each one bearing a strong branch or a dormant bud. The cuttings are then covered with wet sacking, and transported to the nursery.

Nursery beds are best prepared from soil that has been cultivated for many years and is free from cockchafer larvae, which could quickly destroy the new roots. Heavy soils are preferable as they will retain water more effectively. Shading must be provided over the beds. Termites should be eliminated.

Cuttings are set in the soil so that the culm ends and the branch bases are just below the soil level. If the branch base has more buds on one side than the other, the side with most buds should face downwards. Downward-facing buds are more likely to give rooted shoots.

The culm cuttings have enough reserves to support shoot growth for 2-3 months. After that the shoots will start to die back, but roots should just be beginning to develop from the bases of the new shoots. Much larger shoots will then arise from the cutting beds. This second generation of shoots will have abundant rooting, and will form the planting material, once it has hardened off and developed its own branches and leaves. The time required for production of reliable planting material varies from 6 months to 2 years. Most plants will be suitable for planting in the second monsoon after the cuttings were taken. After lifting, the plants should be kept in a shaded area, and watered occasionally until they are planted.

The use of rooting hormones to improve the success of culm cuttings is sometimes advocated. The benefits have not yet been proven in a statistically valid trial, and other factors such as selection of sound material, timely setting of cuttings, and maintaining good environmental conditions are probably more important in most cases.

OTHER PROPAGATION TECHNIQUES

A technique recommended in China uses whole culms buried horizontally, with their rhizomes still attached. This technique produces rooting plants all along the culm in China. Each plant develops from shoots growing from branch buds. In a trial of this technique in Nepal very few plants were produced, even with frequent watering above the rhizome, and with notches cut above each node. This is probably because of the Himalayan spring drought. Branch buds stayed dormant for several years, and shoots only grew from the rhizomes.

A similar technique has sometimes been recommended in India. Holes are cut into each internode of a horizontally buried culm, and the internodes are filled with water. Although this may seem to be a good idea, the interior lining of bamboo internodes is almost entirely waterproof, so this water cannot be transported to the buds and shoots.

In *Bambusa multiplex* small rooting rhizomes are often produced in the air from branch bases. These offsets will develop into successful plants if removed and planted, but only during the monsoon. Similar offsets are produced on *Bambusa nutans* subsp. *cupulata,* but attempts to propagate from them have only been made in the spring, and they were not successful.

Tissue culture has not yet been successful without seed to produce embryogenic callus. Once callus is available however, planting material can be produced indefinitely. Transporting plants from a central laboratory to planting sites is feasible in the plains.

FIELD KEY TO BAMBOO GENERA FOR USE IN BHUTAN

Clump-forming bamboos, culms growing in separate clumps of more than 10 culms:-

Maximum culm diameter > 7cm:-

Culm covered with dark or thick fur, central branches
varied, often very large .... 1. *Dendrocalamus*
Culm with light covering of pale wax, central branches
fairly uniform, usually quite small .... 2. *Bambusa*

Maximum culm diameter < 7cm:-

Maximum internode length > 40cm:-

Leaves with cross veins as well as long veins .... 3. *Borinda*
Leaves with only long veins, no cross veins:-

Culm nodes with no collar or with thick
flat even collar .... 4. *Cephalostachyum & Teinostachyum*
Culm nodes with thin projecting wavy collar ...... 5. *Ampelocalamus*

Maximum internode length < 40cm:-

Buds tall, chilli-shaped .... 6. *Thamnocalamus*
Buds short, onion-shaped:-

Culm sheath blade more than 2cm wide .... 2. *Bambusa*
Culm sheath blade less than 1 cm wide:-

Branches 20 - 70 .... 7. *Drepanostachyum*
Branches 10 - 25 .... 8. *Himalayacalamus*

Spreading bamboos, culms growing separately or in groups of up to 10 culms:-

Culms with rings of thorns around the nodes .... 9. *Chimonobambusa*
Culms with no thorns:-

Leaves with no cross veins, long veins only:-

dbh > 4 cm .... 10. *Melocanna*
dbh 2-4 cm .... 11. *Pseudostachyum*
dbh < 2 cm .... 12. *Neomicrocalamus*

Leaves with distinct cross veins as well as long veins:-

Long rhizome lengths without roots .... 13. *Yushania*
Rhizome rooting at all nodes .... 14. *Arundinaria*

1. DENDROCALAMUS

A genus containing the largest of all bamboo species, forming clumps up to 30m tall. The culms are thin-walled and covered with thick furry wax when young. The branches are usually absent lower down the culm, and are very variable in size, some being more than 5cm in diameter. The bracts at the base of each spherical inflorescence have one ciliate keel (fig. 8), in contrast to those of *Bambusa*, which have two ciliate keels (fig. 12). Most of the species are from subtropical to warm temperate areas, withstanding only a few degrees of frost, although one tropical Malaysian species is planted in the south. All species are easy to propagate by culm cuttings, as the large branches readily produce roots. The thin walls of the culms make the young shoots more liable to attack by shoot-boring larvae, and the dried poles are readily attacked by beetles if not preserved. This genus contains the most important species for edible shoot production in the Himalayas, as well as several general multipurpose species.

fig. 5 - clump appearance

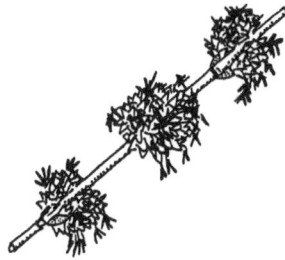

fig. 6 - typical flowers

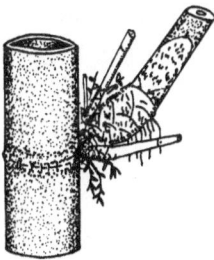

fig. 7 - large branches, thin walls, and furry culm

fig. 8 - bracts at base of inflorescence

KEY TO *DENDROCALAMUS* SPECIES

Culm sheath auricles more than 2cm wide with long spreading bristles . *sikkimensis*
Culm sheath auricles less than 2cm wide:-

    culm sheath blade bent backwards . . . . *giganteus*
    culm sheath blade upright:-

        culm sheath auricles small, rounded, with bristles:-

            culm and culm sheath with dense dark brown
            fur or hairs at first; nodes with many small rootlet buds . . . . . *hookeri*

            culm and culm sheath with light fur or hairs;
            nodes with two rings of white wax and few or no
            small rootlet buds . . . . see *Bambusa clavata*

        culm sheath auricles absent or small, triangular and naked:-

            culms remaining dull with persistent furry wax; culm
            sheaths triangular and naked; leaf sheaths with white hairs:-

                many curving branchlets at culm nodes,
                leaf sheath ligule always long . . . . *hamiltonii* var. *hamiltonii*

                few curving branchlets at culm nodes,
                leaf sheath ligule short or long . . . . *hamiltonii* var. *edulis*

            culms becoming shiny, furry wax deciduous; culm
            sheath auricles absent; leaf sheaths with brown hairs
                . . . . see *Bambusa balcooa*

Dendrocalamus giganteus (Nep. *dhungre bans, rachhasi bans*)   D30

**Culm &
culm sheath**

- broad culm,
  pale wax

- reflexed blade

- hairs few,
  scattered

Leaf sheath

- ligule tall
  & truncate

- no
  bristles

- no
  hairs

Culm sheath ligule & auricles

- tall serrated ligule

- auricles joining base
  of blade, with no bristles

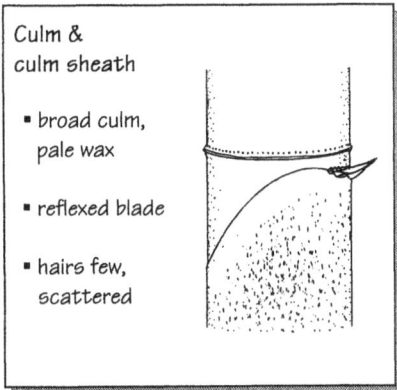

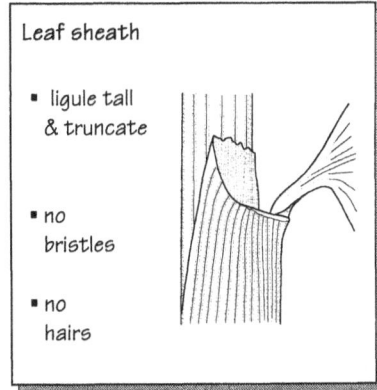

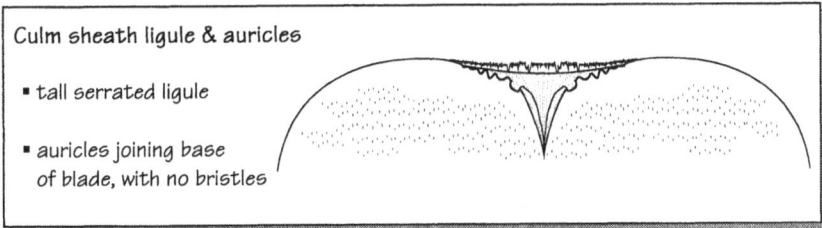

This is the largest bamboo in South Asia, with a maximum dbh of more than 30cm. The tallest culms reach 30m in height. It is similar to the slightly smaller *D. sikkimensis*. The clumps of both species have horizontal yellow culm sheath blades on the new shoots. It can be distinguished from *D. sikkimensis* by the absence of bristles on the auricles of the culm sheaths or leaf sheaths. The hairs on the culm sheath are also much fewer, lighter in colour, and are flattened against the sheath. The reflexed culm sheath blades distinguish it from *Bambusa balcooa*.

The leaf sheaths become quite red at the tips, and it has long pendulous flowering branches. The large diameter culms are used as pillars or for making storage containers, and for special uses such as road barriers. However, they are too large for most general purposes and this species is not widely cultivated. The very large leaves are used as animal fodder.

This species is found across the plains of West Bengal and Assam, and along the edge of the Bhutanese hills. It is a tropical species from Malaysia, and may not grow well above 1,000m.

Propagation of this species is not easy. The large size of the rhizomes makes it difficult to use the traditional technique. Culm cuttings would be successful as the branches are large but there are few branches in the lower part of the culm.

13

Dendrocalamus hamiltonii (Nep. *choya, tama, ban bans,* Dz. *patsa*) D4/D46
var. **hamiltonii** and var. **edulis**

Culm &
culm sheath

- brown/white
  fur

- curving
  branchlets

- auricle with
  no bristles

Leaf sheath

- tall rounded
  ligule

- no auricle
  or bristles

- white
  hairs

Culm sheath ligule

- rising steeply in centre

- margin mostly serrate,
  dentate at edges

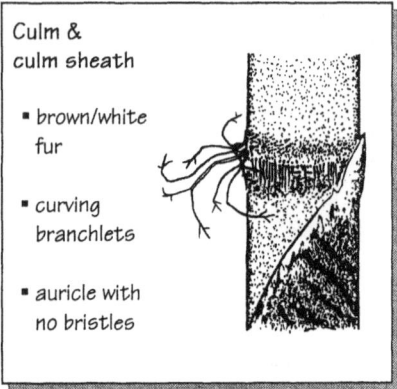

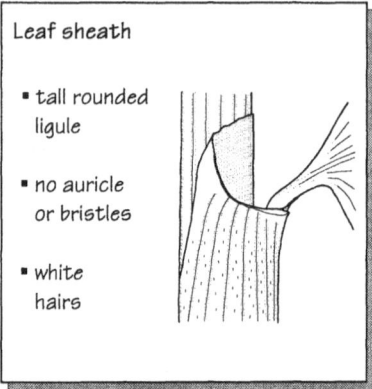

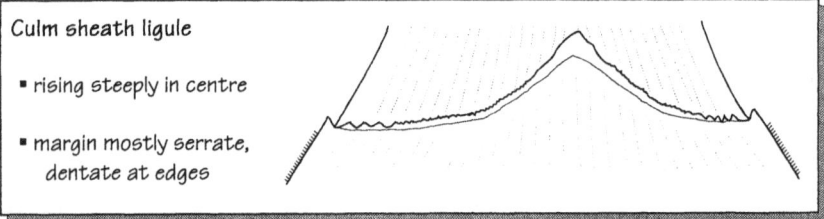

The most common bamboo of the subtropical forest along the outermost foothills of the entire Himalayan range, often cultivated further into the hills.

D. hamiltonii has a long leaf sheath ligule, naked triangular auricles on the culm sheaths, persistent pale fur on the culms and long drooping culm tips.

Var. *hamiltonii* has small red flowers. Var. *edulis* is found east of Darjeeling. It has larger yellow flowers, fewer curved branchlets at the culm nodes, shorter leaf sheath ligules, more pendulous culm tips, and sweeter shoots.

The culms are thin-walled and very flexible, giving the best weaving material of all large bamboos, but the large branches of many varieties make the culms difficult to split. This species is commonly managed without cutting mature culms. New shoots are removed for human consumption, and the large branches are cut for weaving material and fodder. This often leads to tightly congested clumps.

Propagation by vegetative means is easy because of the large branches and prolific aerial rooting. Culm cuttings give up to 90% success rates. Small areas of flowering bamboo can be found in most years, and seed is often available.

The combination of multiple uses and ease of propagation by seed or cuttings makes this a highly suitable species for all planting programmes. It also has potential for large scale edible shoot production.

Dendrocalamus hookeri (Dz. *pagshi,* Nep. *bhalu, kalo bans*)

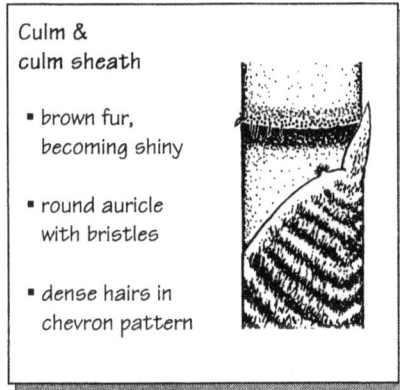

Culm &
culm sheath

- brown fur,
  becoming shiny

- round auricle
  with bristles

- dense hairs in
  chevron pattern

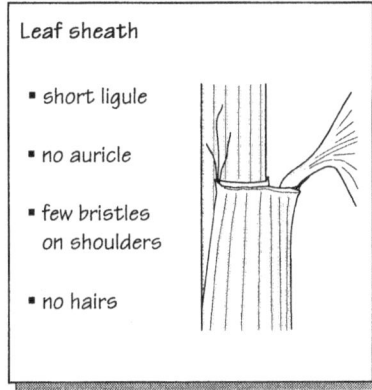

Leaf sheath

- short ligule

- no auricle

- few bristles
  on shoulders

- no hairs

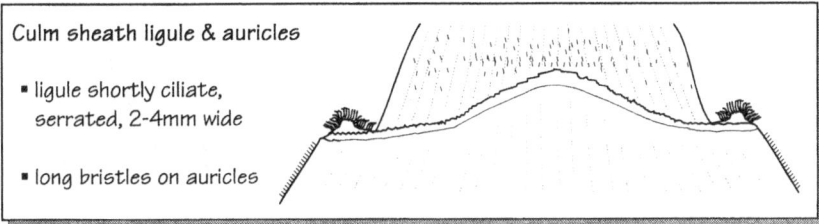

Culm sheath ligule & auricles

- ligule shortly ciliate,
  serrated, 2-4mm wide

- long bristles on auricles

An infrequent species of southern and eastern Bhutan, similar to *D. sikkimensis.* Both species have dense furry wax on the culms and thick brown hairs on the culm sheaths. These hairs are not so dense in *D. hookeri,* and are often in a distinctive chevron pattern where they have rubbed off during growth.

It can be distinguished from *D. sikkimensis* by the much smaller auricles on the culm sheaths, and by the leaf sheaths which have fewer bristles. *D. hamiltonii* is similar, but has no bristles on the culm sheath auricles and longer leaf sheath ligules. *Bambusa clavata* is also similar but has lighter culm sheath hairs and a wider culm sheath ligule.

The culms can reach a maximum diameter of 16cm and a top height of 25m when unthinned, but they are usually 8-9cm in diameter and 15-18m tall. Culm walls are thin but not very flexible, so it is used for general construction and fencing, rather than weaving. The leaves are large and can be an important fodder source in winter. Sections of larger culms are often used as containers. Although this species appears very similar to *D. asper,* which is widely grown for its edible shoots, the shoots of *D. hookeri* are bitter.

Propagation by all the vegetative techniques is easy because of the abundance of aerial roots and very strong branching. This species can even be propagated from the bases of the large branches on their own.

Found from 1,200-2,000m.

Dendrocalamus sikkimensis (Nep. *dhungre bans*, Shar. *dem chherring*) D48

Culm &
culm sheath

- dense brown
  furry wax

- large auricles
  with bristles

- dense upright
  hairs

Leaf sheath

- short
  ligule

- long
  bristles

- no
  hairs

Culm sheath ligule

- margin curled & feathered

- large auricles with
  long spreading bristles

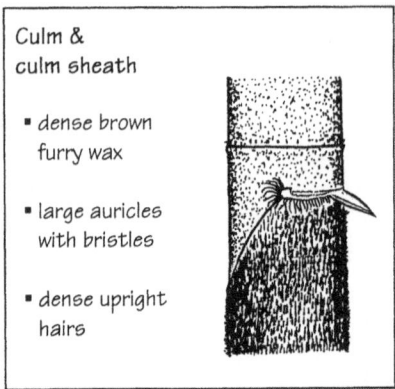

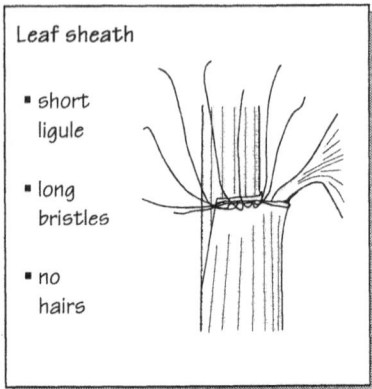

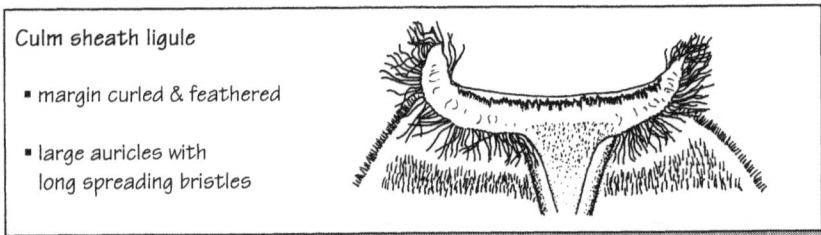

Similar to *Dendrocalamus hookeri*, but taller. The brown hairs on the culm sheaths are denser, more erect, and more persistent. The auricles on the culm sheaths are much larger and wider, and have more bristles. It is also similar to *Dendrocalamus giganteus*. Both species grow very large and have horizontal culm sheath blades on the new shoots. However, this species is smaller, and can easily be distinguished from *D. giganteus* by the long bristles on the auricles of the culm sheaths, and by the tough smooth leaf sheaths.

The culms grow to a very large size and sections are used for carrying water and as general containers. Culm walls are thin and not flexible enough for weaving. According to Gamble (1896) the foliage is said to be poisonous to cattle and horses, and the shoots are very bitter.

Propagation by vegetative techniques is easy because of the abundance of aerial roots on the strong branches. Culm cuttings would be very successful with this species. Seed has not been collected.

Further information on distribution is required. It grows near Tingtibi on the Mangde Chhu in Shemgang district, and it is reported to occur widely in eastern Bhutan. It is common in Sikkim from 1,100-1,900m.

2. BAMBUSA

A genus containing large bamboos up to 26m in height, as well as several smaller species less than 12m in height. These clump-forming bamboos are similar to *Dendrocalamus* species, but they are generally smaller, with straighter culms and thicker culm walls. The leaves are smaller, and the new culms usually have a thin pale waxy covering rather than dense furry wax. The branches are more uniform in size. Central branches are usually less than 5cm in diameter (fig. 11), so that propagation from culm cuttings can be difficult. Branches are often found right to the base of the culm, and they are thorny in some species. The flowers are in spiky inflorescences (fig. 10 cf. fig. 6), and the bracts at the inflorescence base have two ciliate keels (fig. 12 cf. fig. 8). The species are tropical to subtropical in distribution, occurring up to 1,600m. They provide a very important source of construction material in warmer areas of southern and eastern Bhutan, and are also used to make bows for archery.

fig. 9 - clump appearance

fig. 10 - typical flowers

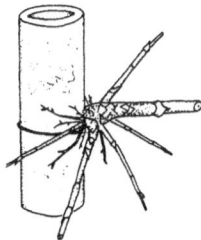

fig. 11 - small branches, thick walls, and glossy culm

fig. 12- bract at base of inflorescence

KEY TO *BAMBUSA* SPECIES

Maximum culm height less than 10m, no hairs on new culm sheaths .... *alamii*

Maximum culm height more than 10m, new culm sheaths densely hairy:-

    Culm sheath auricles large, more than 1.5cm in width or height:-

        culm sheath auricles slightly pointed, often curling
backwards; edges of culm sheath blade bearing long
bristles towards the base .... *vulgaris*

        culm sheath auricles rounded, upright; edges of culm sheath
blade sometimes bearing a few bristles:-

            culm sheath blade prominently cupped, persistent;
both culm sheath auricles wider than their height;
bases of culms without yellow stripes; culms up to
25m tall, very straight .... *nutans* subsp. *cupulata*

            culm sheath blade not prominently cupped, deciduous;
one culm sheath auricle rounded, taller than its width;
bases of some culms with faint yellow stripes; culms
up to 15m tall, slightly crooked .... *tulda*

    Culm sheath auricles absent or small, less than 1.5cm in width and height:-

        no auricles, even on new culm sheaths; leaf sheaths with
brown hairs .... *balcooa*

        culm sheath auricles small, rounded; leaf sheaths with no hairs ... *clavata*

Bambusa alamii (Nep. *mugi bans*) B42

Culm &
culm sheath

- long narrow
  internodes

- asymmetrical
  auricles with
  long bristles

- no hairs

Leaf sheath

- long bristles

- large
  auricles

- short
  ligule

- no hairs

Culm sheath ligule & auricles

- blade separated from auricles

- ligule 1-2 mm wide with
  no serrations on margin

- one long extended auricle,
  one small rounded auricle

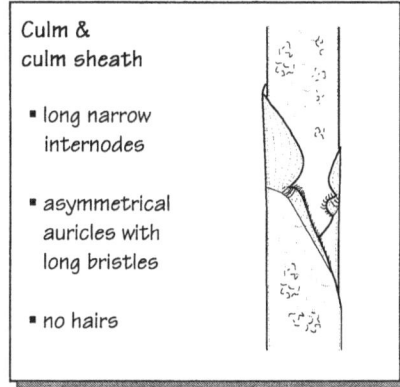

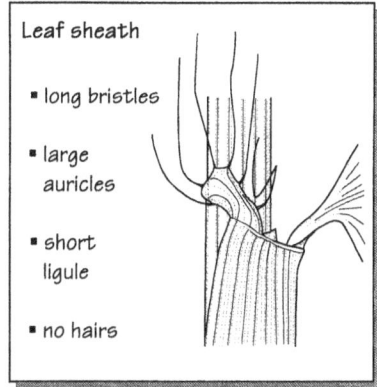

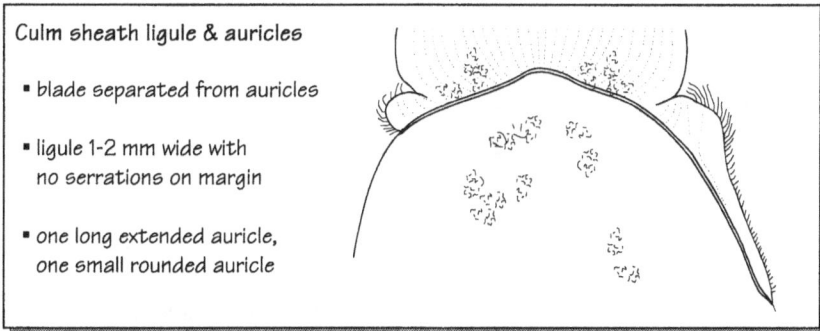

This species is similar to Chinese Hedge Bamboo, *Bambusa multiplex*, and is widely cultivated in Bangladesh. It is sometimes found near Bhutan's southern border.

Like *B. multiplex* it reaches a maximum diameter of around 4cm and is short for a *Bambusa* species, with a maximum height of 10m. It has very straight culms with long internodes, and little swelling at the nodes.

This species is recognised by its small size, and by its glabrous asymmetrical culm sheaths with large auricles bearing long bristles. The auricles are separated from the culm sheath blade, and one auricle can extend almost half-way down the sheath. The leaves have no hairs, and the culm sheath is firmly attached to the culm below the branch bud.

The narrow culms with long straight internodes and small branches are highly suitable for splitting into weaving strips. In the heat of the plains small bamboos from genera such as *Drepanostachyum* and *Himalayacalamus* will not grow, and species such as this are a very useful substitute for the smaller hill bamboos.

It is not known whether this species is native to South Asia, or whether it has been introduced from China in the past.

Bambusa balcooa   (Nep. *dhanu bans, ban bans,* Dz. *jhushing*)   D23

Culm & culm sheath
▪ brown fur at first, shiny later
▪ thorny branchlets
▪ no auricles

Leaf sheath
▪ a few short bristles
▪ short ligule
▪ brown hairs

Culm sheath ligule
▪ blade edges wavy
▪ ligule margin wavy, finely serrated

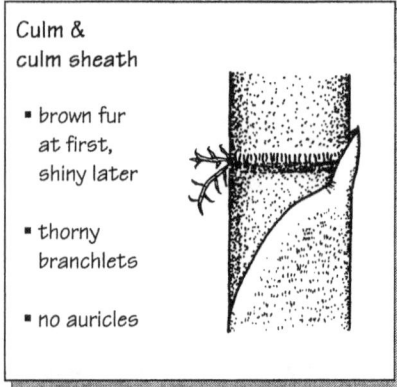

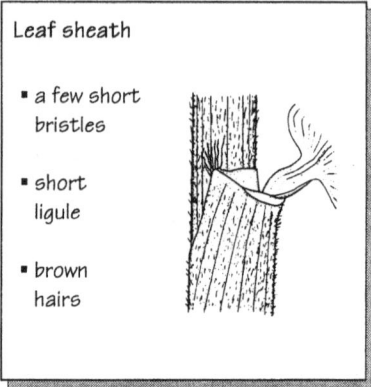

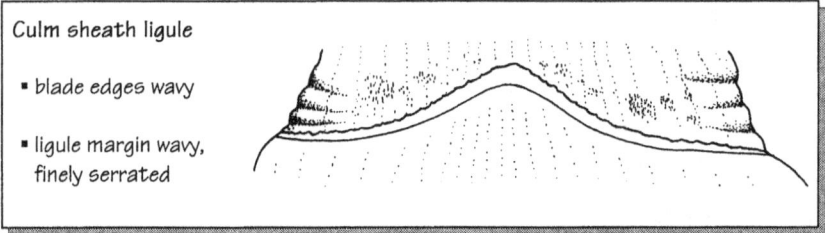

This is a large thick-walled bamboo with strong branching and thorn-like branchlets lower down the culm. It can reach a diameter of 16cm and a height of up to 25m. It is similar to species of *Dendrocalamus*, having thick furry culm wax, densely hairy culm sheaths, and large branches. It is easy to recognise because of the brown hairs on the leaf sheaths and the small curving thorn-like branchlets. The thorns are smaller than those of *B. bambos*, and there are fewer hairs inside the culm sheath blade. It can be separated from all other large bamboos in the country by the absence of auricles on the culm sheaths.

The poles are highly valued in India. They are an important raw material which can be marketed in large quantities, as scaffolding and for weaving into panels used for low-cost housing. They are generally a little too large for village use, and the heavy branching makes them difficult to split by hand. They are reserved for a few uses such as making pillars and beams.

This is an adaptable species, widespread across West Bengal and Assam, and it grows well from Calcutta up to around 1,600m. It tolerates drier conditions better than many bamboos, but can suffer from the bamboo blight syndrome on poorer sites.

The large size of this species along with its thorny branchlets make it a good choice for slope stabilisation. The large branches make it easy to propagate from culm cuttings.

Bambusa clavata   (Nep. *chile bans*, Dz. *jhushing*)   D41

Culm &
culm sheath

- white bands
  at nodes

- small auricle,
  short bristles

- dark brown
  hairs

Leaf sheath

- a few long
  bristles

- short cilate
  ligule

- no hairs

Culm sheath ligule & auricles

- blade deciduous

- very tall feathered ligule
  with large single cleavage

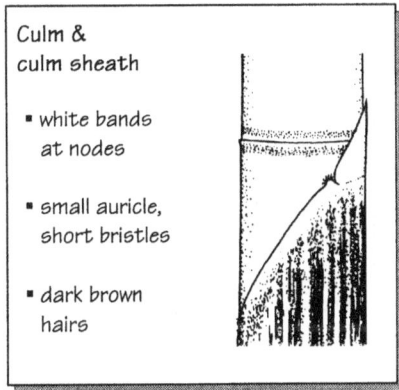

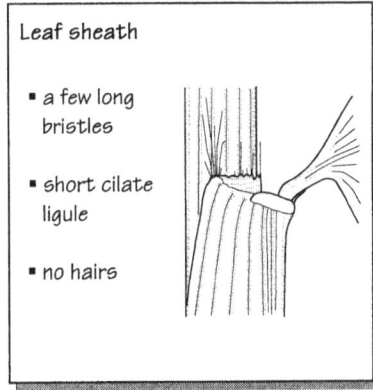

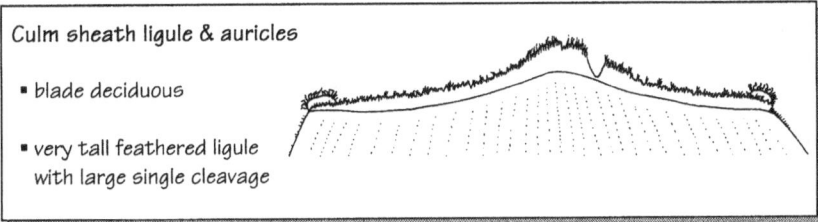

A common species in southern and central Bhutan with thin-walled upright culms that are used for general constructional purposes.

This species can be recognised by the distinctive culm sheath ligule on larger culms. The ligule is very tall with a feathered margin, and it often has a single wide cleavage to one side of the centre. The culm sheaths have small auricles with short bristles, and dark brown hairs. The new culm sheaths are yellow and spear-like with closely pressed culm sheath blades, and hairs arranged in prominent vertical dark brown streaks.

There are persistent white bands above and below the culm nodes. The culms reach a maximum diameter of

9cm and a top height of 18m and are used for light construction, and also for low quality weaving. The small branches and unraised nodes allow easy splitting. The leaves, although small, are used for fodder and the shoots are sometimes eaten.

Despite the lack of branching lower down the culm this species does produce aerial roots on the central branches, and the culm cutting technique of vegetative propagation should be successful.

This species is found from 300m to 1,600m. A high proportion of the clumps flowered gregariously in southern Bhutan from 1986 to 1989. The clumps died, but some new clumps regenerated from seedlings.

Bambusa nutans subsp. **cupulata** (Nep. *mal bans*, Dz. *jhushing*)   B1

Culm & culm sheath	Leaf sheath
• strongly cupped deciduous blade	• a few tall bristles
• broad auricles, reddish bristles	• small auricle
• jet-black hairs	• short ligule
	• no hairs

Culm sheath ligule

• ligule 1-2mm tall

• margin finely serrated

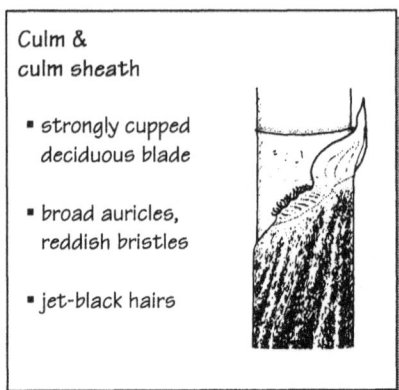

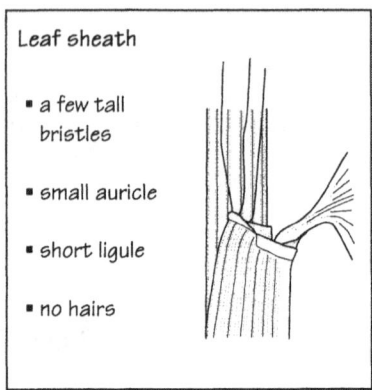

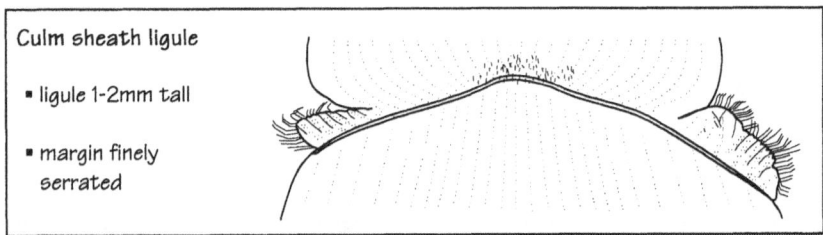

The commonest cultivated bamboo in the lower hills of Bhutan from 300m to 1,500m. It is very similar to *Bambusa tulda*, both species having large wide culm sheath auricles, but it is taller and straighter, and has smaller branches. This subspecies can be recognised by the strongly cupped culm sheath blades, which are deciduous. The hairs on the culm sheath are blacker than those of similar bamboos.

The poles reach a maximum diameter of 10cm and are up to 23m long. They are strong and highly prized for all constructional purposes. They can also be used for the weaving of rough baskets and mats, as the branches are small and the poles split easily. The poles are very straight and the unraised nodes with small branches and small leaves make it a very attractive and clean-looking bamboo. Leaves are widely used for fodder. The shoots are bitter and are not eaten.

This is one of the most desirable bamboos for many end-uses, because of its long straight culms and small branches. It will tolerate dry stony sites quite well, and it is common from the eastern Himalayas to the Bay of Bengal.

This subspecies does not generally root well from culm cuttings as the branches are small and there are few aerial roots. However, at lower altitudes there are forms with shorter culms and larger branches which can be more successful. It is easiest to establish this subspecies by the traditional technique.

Bambusa tulda   (Nep. *singhane bans*, Dz. *jhushing*)   B22

Culm &
culm sheath

- faint yellow
  stripes

- tall rounded
  auricle

- dark brown
  hairs

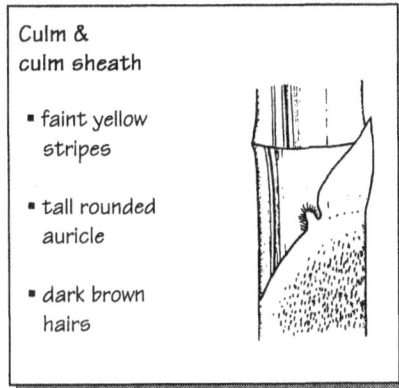

Leaf sheath

- auricle with
  bristles

- short
  truncate
  ligule

- no hairs

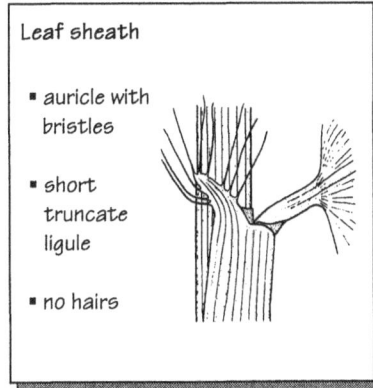

Culm sheath ligule & auricles

- broad and tall auricles

- ligule 1-3mm tall,
  margin finely serrated

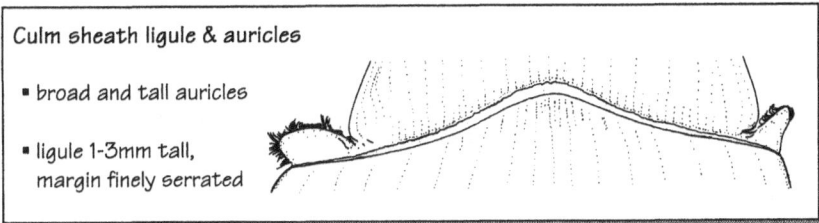

A species which is usually rare in the Himalayas, but is common in Chirang district. It has strong very upright culms, but they are quite short and can be rather crooked, with swollen nodes and very heavy branching.

It can be distinguished by the prominent leaf sheath auricles, and by the culm sheath auricles, one of which is quite tall, while the other is much broader. There are often faint yellow stripes on the lower internodes of some culms. Unlike *B. nutans* subsp. *cupulata,* the culm sheath blade is not strongly cupped and it is more persistent. The hairs on the culm sheath are brown rather than black. It is similar to *B. vulgaris,* which also has varieties with striped culms. It can be distinguished from that species by the absence of bristles on the lower edges of the culm sheath blade and by the larger leaf sheath auricles. It is also shorter.

The poles can reach a maximum diameter of 7cm and a length of 15m, although they are often smaller. They are very thick-walled and used for constructional purposes. Leaves can be used for fodder, but they are small. The shoots are not edible.

The thick walls and strong branching make it easy to propagate this species by any vegetative means, and branches on their own may root successfully.

The ease of propagation makes it an easy species to plant, but *Bambusa nutans* will give longer straighter culms.

Bambusa vulgaris   (Nep. *teli bans*)   B50

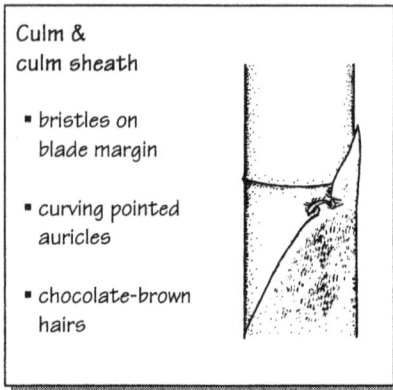

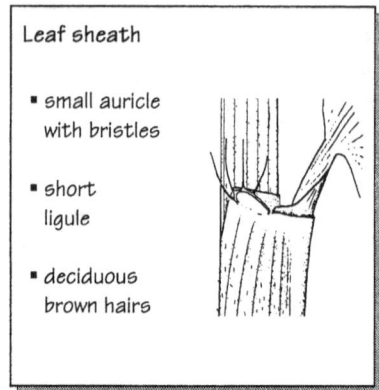

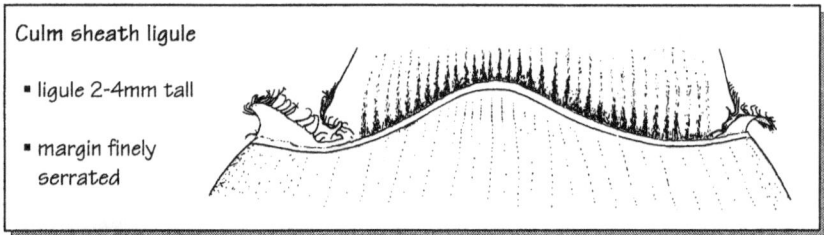

Culm &
culm sheath

- bristles on
  blade margin

- curving pointed
  auricles

- chocolate-brown
  hairs

Leaf sheath

- small auricle
  with bristles

- short
  ligule

- deciduous
  brown hairs

Culm sheath ligule

- ligule 2-4mm tall

- margin finely
  serrated

A large bamboo with strong culms common in the plains of India, but rare in Bhutan.

This species can be recognised by the culm sheath auricles, which are slightly pointed and curve backwards. There are also bristles on the undulating lower edges of the culm sheath blade. The interior of the blade has prominent lines of dense, dark, upright hairs.

It is similar in culm characteristics and uses to other *Bambusa* species such as *Bambusa nutans* or *Bambusa tulda*, having strong thick walls, and it provides a general purpose construction material. The culms are not flexible and are not generally used for weaving. They are straighter than those of *B. tulda* but shorter than those of *B. nutans.*

This species is widely cultivated throughout the tropical world, largely because it can be propagated very easily. Culm cuttings are very successful and the branches will root rapidly on their own in a moist site.

Ornamental varieties of this species with prominently striped yellow and green culms are widely grown in Indian gardens. A cultivar with short swollen nodes, cv. Wamin, is also cultivated, sometimes as a pot-plant.

Cultivated forms of this species do not flower gregariously, and will often remain in vegetative growth indefinitely.

3. BORINDA

Clump-forming frost-hardy bamboos, found in temperate forest from 1,800-3,200m, up to 10m tall, with prominent cross-veins on the leaves, tall buds, and finely-grooved culm internodes that are up to 50cm long. These bamboos are quite similar to *Thamnocalamus* species, but they are larger, with seven branches in the first year, two additional branches extending behind the new culm (fig. 16). The long finely-grooved culm internodes are different to those of *Thamnocalamus* and *Himalayacalamus*

species, which are short and smooth. The culms are very similar to those of *Cephalostachyum latifolium* and also *Ampelocalamus patellaris*, but *Borinda* species have taller buds and cross-veined leaves. The rhizomes are less than 30cm long, shorter than those of the two temperate spreading genera *Yushania* and *Arundinaria*, fig. 15. The culms are flexible and easily split, and they are highly valued for production of weaving material. Tree seedlings can easily regenerate between the clumps.

fig. 13 - clump appearance

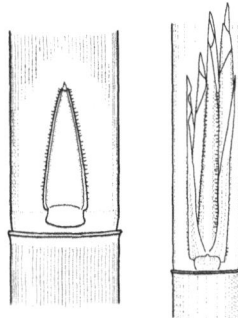

fig. 14 - tall buds, grooved culm

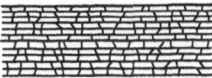

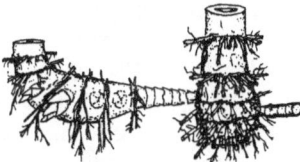

fig. 15 - leaf veins and rhizome

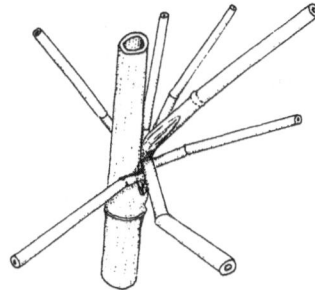

fig. 16 - young branches

Borinda grossa   (Dz. *rhui, baa*)   T43

Culm & culm sheath

- finely-grooved waxy culm

- no auricle, upright bristles

- blade deciduous

Leaf sheath

- hairy leaves

- upright bristles

- no auricle or hairs

Culm sheath ligule

- tall smooth twisted bristles

- ligule tall, ciliate, concave, & hairy

- long narrow blade

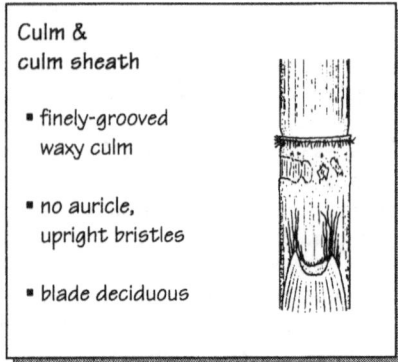

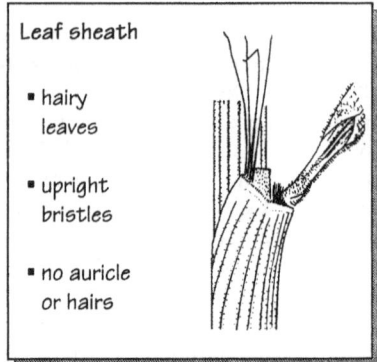

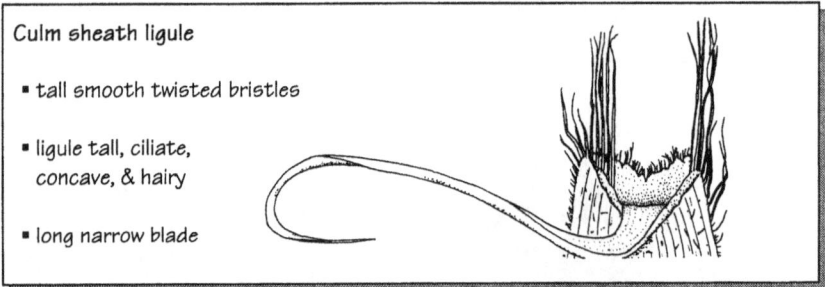

An important species, widely found between 2,600m and 3,200m, often in association with hemlock. It is naturally restricted to wetter temperate mixed coniferous forest areas of central and eastern Bhutan, but is also cultivated.

This species reaches large dimensions for the altitude at which it grows. It has culms up to 10m tall and up to 4.5cm in diameter, and leaves up to 25cm long. It is easily distinguished from the other frost-hardy bamboos by its large finely-grooved culms which are a pale blue colour. *Yushania hirsuta* is similar, but has rough, rather than grooved culms, and large auricles, and long rhizomes so that it does not grow in clumps. Three similar species have been described from Tibet, and one from Nepal.

Because of its level nodes, thin walls, and long internodes, this bamboo is easily split into strips and woven into mats for house-roofing. It is also used for making fence sections.

It is possibly one of the most important minor forest products in central Bhutan. It is widely harvested throughout its range, and is often seen being dragged along the road in bundles, or being split and woven into fence sections or roofing mats at the roadside. It is also planted around houses and villages in central Bhutan.

Propagation of this species is only possible using the traditional technique, and long poles should be used to allow early growth of branches and leaves.

4. CEPHALOSTACHYUM & TEINOSTACHYUM

Clump-forming tropical and sub-tropical bamboos up to 10m tall and up to 5cm in diameter, found in high rainfall forests. Smaller than *Bambusa* and *Dendrocalamus* species but larger than temperate bamboos, these two closely related genera are useful for weaving into mats, having flexible culms with internodes of up to 1m. The long internodes, often with fine ridges, are similar to those of *Borinda grossa,* but the leaves only have long parallel veins (fig. 20), without any of the short cross-veins seen in frost-hardy bamboos, (fig. 15). They are also similar to those of *Ampelocalamus patellaris,* but without the frilly collars at the culm nodes, or the long-fringed culm sheath margins. The buds are short and rounded, (fig. 19), while those of the frost-hardy genera are tall and narrow, (fig. 14). The tips of the culms are long and thin, and may hang down to the ground, or sprawl over tree branches. Culm sections can be made into flutes. They are mainly found in natural forest, and are rarely cultivated.

fig. 17 - clump appearance

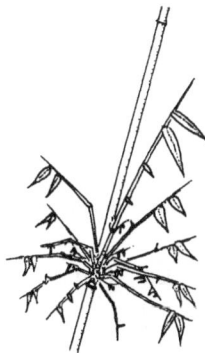

fig. 18 - long internodes and spreading branches

fig. 19 - mid-culm bud

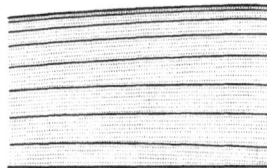

fig. 20 - parallel leaf veins

Cephalostachyum latifolium   (Dz. *jhi* Nep. *ghopi bans*)   U43

Culm &
culm sheath

- thick hairy nodes

- ring of
  light wax

- culm ridged
  and scabrous

Leaf sheath

- tall flattened
  white bristles

- ligule long
  and glabrous

- sheath edge thin
  with cross-veins

Culm sheath apex (interior)

- tall flattened white bristles

- raised delicate shoulders
  with visible cross-veins

- ligule very short, blade
  with dense brown hairs

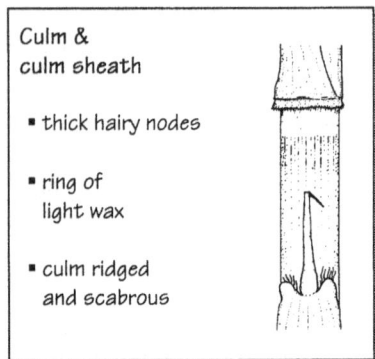

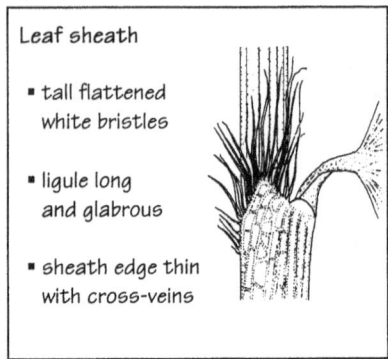

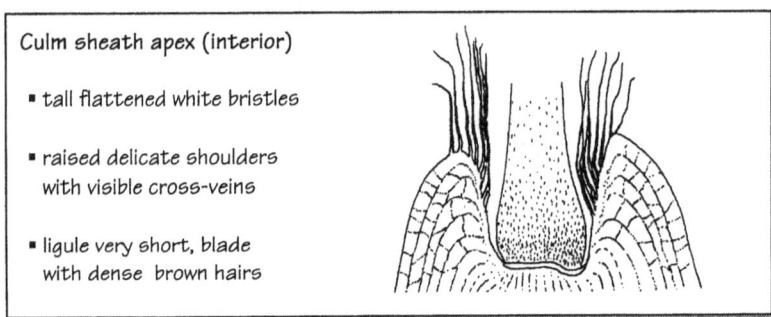

A distinctive species of the cooler subtropical forests of western and southern Bhutan, usually found between 1,500m and 2,000m. Culms are up to 15m long, and up to 5cm in diameter, with internodes up to 1m in length. The straggling clumps have long pendulous culm tips, and very large leaves for the size of the culms.

The ridged culm sheaths with thin edges and tall shoulders distinguish this from the similar species *Teinostachyum dullooa.* The leaf sheaths also have tall shoulders, and both culm and leaf sheaths have long white bristles when young. The bristles are delicate and deciduous, leaving hardly any trace once they have fallen. Culm nodes and sheath bases have short light-brown hairs. The culm nodes are swollen, with a corky collar similar to that of *Ampelocalamus patellaris,* but much thinner.

The flexible culms with very long internodes are highly suitable for weaving, and the broad leaves, up to 30cm long and 6cm wide, make excellent animal fodder. This bamboo is harvested from the forest on a regular basis. Most of the clumps of this species near Phuntsholing flowered in recent years, and many seedlings and small young clumps can now be found in the forest. Some of the orange flowers were collected for use as paint brushes.

Teinostachyum dullooa   (Nep. *tokhre bans*)   U42

Culm &
culm sheath

- level nodes
  with no hairs

- ring of
  light wax

- culm smooth

Leaf sheath

- tall rounded
  red bristles

- ligule short
  and hairy

- sheath edge
  tough

Culm sheath apex (exterior)

- tall rounded copper-coloured bristles

- level tough shoulders
  with no visible cross-veins

- ligule very short, blade
  with dense brown hairs

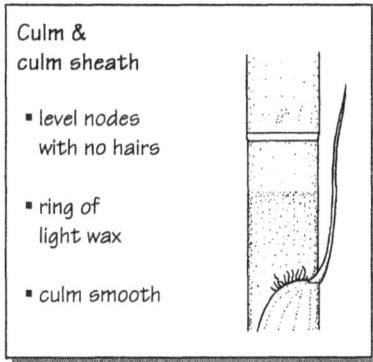

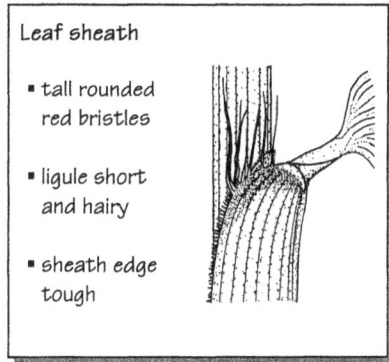

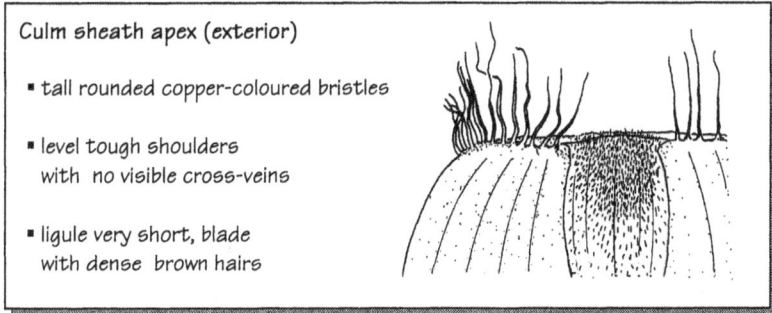

A species of southern and eastern Bhutan, found in subtropical forests from 500m to 1,500m, reaching a height of 10m, and a diameter of 4cm.

The straggling clumps look similar to those of *Cephalostachyum latifolium*, but the leaves are much smaller. The culm sheaths are smoother with flat level shoulders. The shoulders and edges are tough, and there are copper-coloured rounded bristles on the shoulders. The leaf sheaths are also tough, without the cross-veins seen on the leaf sheaths of *Cephalostachyum latifolium*. The leaf sheath shoulders are rounded rather than tall, with fewer, darker bristles which merge with the cilia down one edge. The leaf sheath ligules are shorter,

and more densely hairy. The culm nodes are level, not swollen, and the sheath bases have no hairs. The culm internodes are smooth with no ridges, and they are very shiny at the base.

The flexible poles with their long internodes are very useful for weaving, although they are thin-walled, and not very strong. The walls are thinner than those of *Cephalostachyum latifolium* but thicker than those of *Pseudostachyum polymorphum*.

This species is not usually cultivated, although it is harvested from the forest. It also occurs through Assam, Burma and Bangladesh, and is related to species of *Schizostachyum* from S.E. Asia.

5. AMPELOCALAMUS

Clump-forming thornless bamboos, up to 12m tall, cultivated from 1,200m to 2,000m, with long internodes, no cross-veins on the leaves, and short buds. This genus is very similar in appearance to two other medium-stature subtropical clump-forming genera, *Cephalostachyum* and *Teinostachyum*, although their branches and inflorescences show that they are not closely related at all. *Ampelocalamus* species have finely ridged culm internodes of more than 40cm, short broad buds, no cross-veins on their leaves, and long pendulous culm tips. They can be distinguished from *Cephalostachyum* and *Teinostachyum* species by their distinctive branching. The branches are all similar in size, and are arranged in vertical groups. They curve outwards from the culm and have swollen bases, and the larger branches often bear aerial roots. Most species are found only in China, but one species extends from Yunnan along the Himalayas to central Nepal, and is known in Darjeeling district and Sikkim.

fig. 21 - pendulous culms

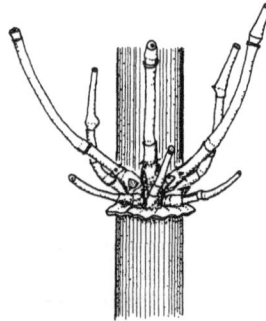

fig. 22 - mid-culm branching

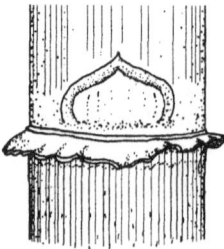

fig. 23 - bud and corky collar

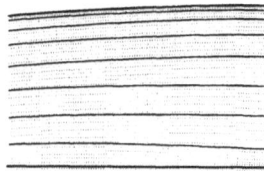

fig. 24 - parallel leaf veins

Ampelocalamus patellaris (Nep. *nibha, ghopi bans*) T3

Culm & culm sheath	Leaf sheath
• finely ridged culm surface	• tall bristles on shoulder
• wavy cork collar	• no auricle
• black hairs on culm	• ligule short with bristles
• feathered edges to culm sheath	• no hairs on sheath

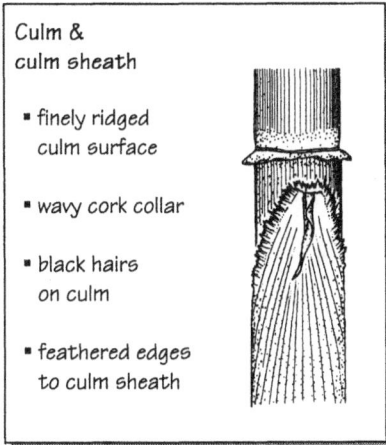

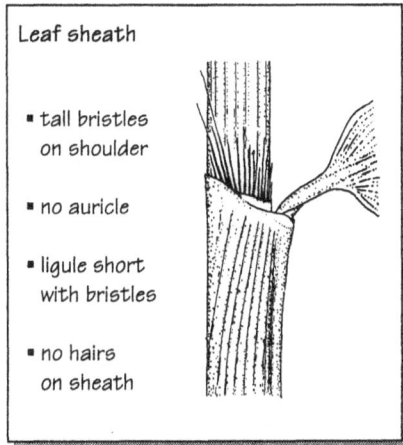

A useful and also very attractive cultivated species, with poles reaching 5cm in diameter and 12m in height. In its natural forest environment this is a scrambling bamboo, but it is usually cultivated in self-supporting clumps.

This is a very easy species to recognise, as the culm sheaths have distinctive long-fringed edges at the top. The leaf sheaths have no auricles, but they have a few very upright bristles, and the edge of the ligule also has long bristles and cilia. The culms also make this species easy to recognise, as they have a distinctive corky collar around each node. This helps to support the flexible upper sections of the culms as they straggle over tree branches.

As it is such a pendulous bamboo, the culms are very flexible, and the long internodes of up to 50cm make them very useful for weaving. This is the main use of this bamboo, as the culm walls are too thin for the culms to be of structural value. The leaves are large, up to 40cm long, and can be used as fodder.

The branches are irregular in shape, with curving internodes, and swollen nodes. This allows re-orientation towards the light, and helps to support the scrambling branches. The central branch is only slightly larger than the rest, and it often bears aerial roots. Propagation by culm cuttings should be feasible because of these aerial roots.

This species was first described with the name *Dendrocalamus patellaris*, but it is now known that the flowers which were originally collected came from a clump of *Dendrocalamus hamiltonii* instead. It occurs from 1,200 to 1,800m, and is likely to be encountered in Samchi district. A widespread flowering occurred around 1980.

6. THAMNOCALAMUS

Clump-forming thornless frost-hardy bamboos, up to 5m tall, found from 2,800m to 3,500m in mixed coniferous forest, with cross-veined leaves, smooth or waxy culms, few branches, and usually with upright culm sheath blades. These are the highest altitude clump-forming bamboos, growing well above the range of all the species of *Cephalostachyum, Teinostachyum, Ampelocalamus,* and also *Drepanostachyum.* They can always be distinguished from those genera by the cross-veins on their leaves (fig. 27). In distinction to *Borinda* the branches are fewer, and do not extend behind the culm in the first year (figs. 26 and 16). The culm internodes are also smoother and shorter. Culm buds are taller than their width (fig. 28), unlike the short buds of *Himalayacalamus* species. The rhizomes are solid, and shorter than those of *Yushania* and *Arundinaria,* less than 30cm long (fig. 27). The culms are small and brittle and not widely used, but the shoots and leaves provide important food and cover for wildlife.

fig. 25 - clump appearance

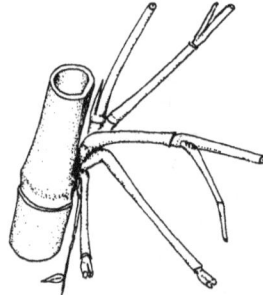

fig. 26 - typical branching

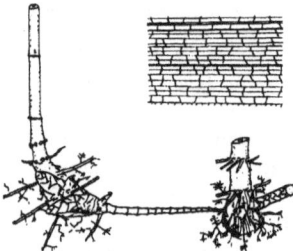

fig. 27 - rhizome and leaf veins

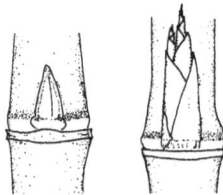

fig. 28 - tall buds, smooth culms

Thamnocalamus spathiflorus   (Dz. *hum*, Nep. *rato nigalo*)   T52, T44

Culm &
culm sheath

- smooth culm
  ± waxy layer

- upright blade

- dense hairs
  at first

Leaf sheath

- spreading
  bristles

- dark petiole

- ciliate callus

- no hairs

Culm sheath apex
subsp. *spathiflorus*

- sides nearly
  symmetrical

- flat ligule

Culm sheath apex
var. *bhutanensis*

- sides strongly
  asymmetrical

- rising ligule

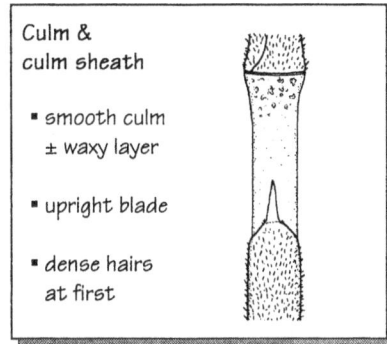

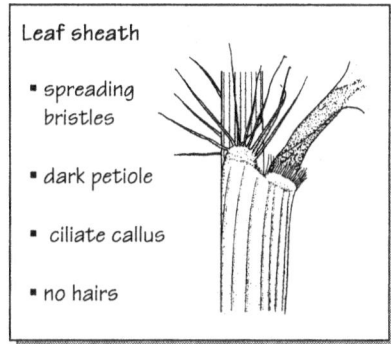

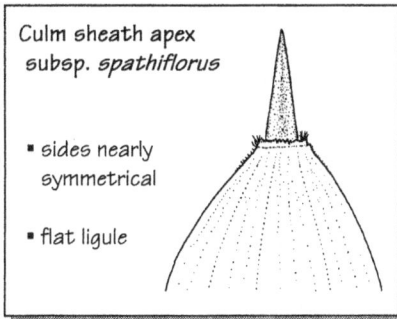

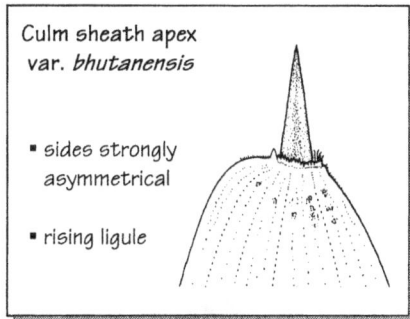

A common species, found between 2,800m and 3,500m, usually above *Borinda grossa* and *Yushania hirsuta* or *Y. pantlingii*, but below flatter areas with widespread *Yushania microphylla* and *Arundinaria racemosa*. It prefers steeply sloping sites.

This species extends right along the Himalayas, and has several subspecies and varieties. It has hairless leaves, often on long pendulous branchlets with many short internodes. The variety *bhutanensis* of central and eastern Bhutan is mainly distinguished from subsp. *spathiflorus* of western Bhutan by its asymmetrical culm sheaths with one horizontal shoulder. It also has irregular auricles, as well as

much thicker wax on the new culms. The culms are smooth and shiny in subsp. *spathiflorus*, becoming yellow or red when exposed. They are waxy in var. *bhutanensis*, becoming yellow or black with age. This species is not commonly harvested, especially when larger bamboos are available nearby, as it has small culms that are brittle and have swollen nodes, which makes them unsuitable for weaving. It is very important for wildlife however, providing food for animals such as red pandas and bears, and shelter for birds such as pheasants. It is also browsed by livestock in winter. It does not hinder regeneration of trees, as seedlings can grow in the gaps between the clumps.

7. DREPANOSTACHYUM

Clump-forming thornless bamboos, up to 5m tall, with many branches, found from 1,000m up to 2,200m in drier subtropical forests, and also cultivated. Leaves have no cross-veins (fig. 30), and culm internodes are less than 40cm long. Branch buds at the nodes of the culms are shorter than their height, and are always open (fig. 31). The buds have many small initials visible, which will produce up to 70 branches at each node, about 25 growing in the first year. The branches are quite uniform in size and spread around the culm (fig. 32). When growing strongly, the upper half of the culm sheaths are very narrow, and the culm sheath ligules are long and ragged. The sheaths are always rough inside at the top, and this distinguishes them from *Himalayacalamus* species. Rhizomes are short and solid, less than 30cm in length, and similar to those of *Thamnocalamus* and *Borinda*. The culms are valuable for weaving, and the foliage is often fed to animals or browsed in the forest. The new shoots are very bitter.

fig. 29 - clump appearance

fig. 30 - flowers and leaf veins

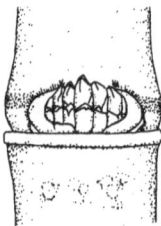

fig. 31 - mid-culm branch bud

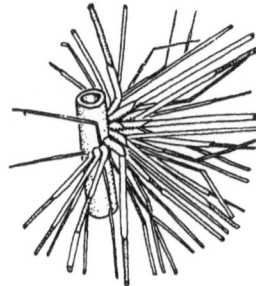

fig. 32 - typical branching

KEY TO *DREPANOSTACHYUM* SPECIES IN BHUTAN

Culm sheath with ring of dense hairs at base .... *annulatum*
Culm sheath without ring of dense hairs at base:-

       leaf sheath hairy with prominent and persistent
       auricles bearing spreading bristles .... *intermedium*

       leaf sheath lightly hairy or glabrous with
       quickly deciduous or absent auricles .... *khasianum*

Drepanostachyum annulatum   (Dz. *him*   Nep. *ban nigalo*)   T47

Culm & culm sheath
• swollen hairy nodes
• dense wax at first
• culm smooth

Leaf sheath
• long ligule
• variable deciduous auricles
• few upright bristles

Culm sheath apex
• narrow shoulders
• ligule long, ragged
• blade narrow, reflexed
• interior densely hairy

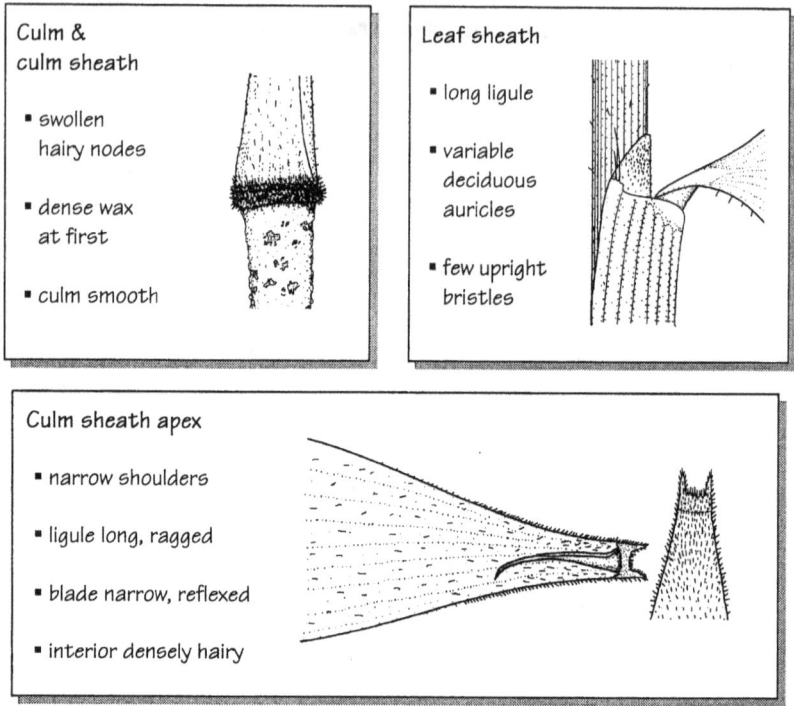

A small forest bamboo, common in Chhukha district from 1,000m to 2,000m, potentially up to 2.5cm in diameter and 5m in height, but usually much smaller because of heavy browsing by animals.

It can easily be recognised by the dense ring of brown hairs at the base of the culm sheath and on the culm node. The young culms also have a dense coating of slippery wax over the entire internode.

The culm sheaths are thin and papery, and on small culms of heavily browsed clumps they are often convex at the shoulders rather than concave. All culm sheaths are rough inside at the top near the ligule, with dense hairs and short spines.

This species provides foliage for browsing livestock, and the culms are collected in small quantities for tying and basket-making. It is also planted around farmland, and it is small enough to be used on the banks of rain-fed terrace risers, where it would be effective in soil stabilisation.

This species flowered near Chhukha in 1987. Most of the clumps died after flowering, but seedlings produced new clumps where there was some natural protection from livestock.

Drepanostachyum intermedium   (Nep. *tite nigalo*)   T1

Culm &
culm sheath

- swollen nodes
  with no hairs

- light wax

- narrow
  sheath apex

Leaf sheath

- long spreading
  bristles

- long ligule

- persistent
  hairs

Culm sheath apex

- sheath with no hairs

- very long ragged ligule

- interior with many
  short spines

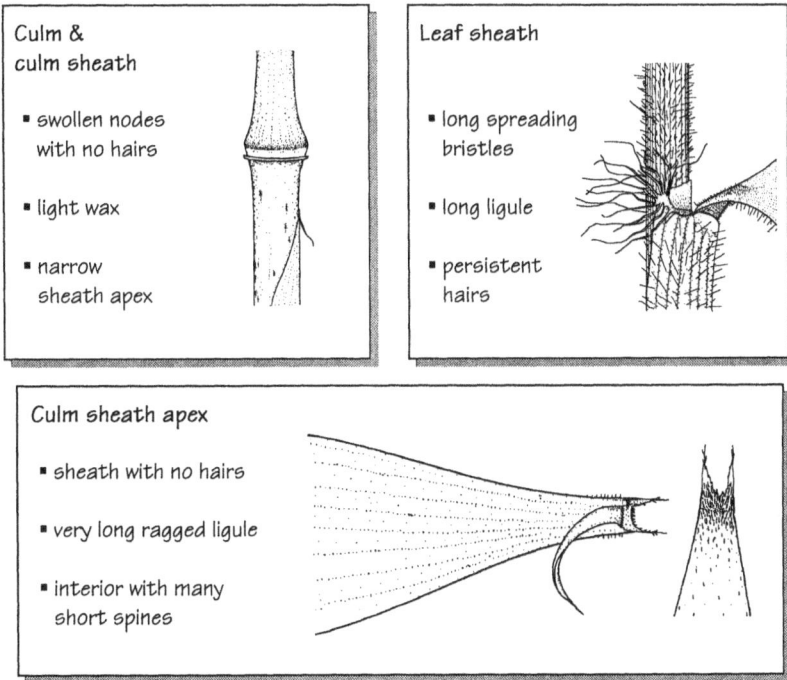

A species found in evergreen oak and chestnut forest, and also cultivated around subtropical farmland. The culms can be up to 2cm in diameter and 4m tall, but they are usually much smaller in the forest because of heavy browsing.

It can be recognised easily by the well developed and persistent leaf sheath auricles, with widely spreading bristles, and by the dense hairs on the leaf sheaths and the undersides of the leaves.

It is found from 1,000m to 2,000m in southern and central Bhutan. When cultivated the main use of this species is basket-making, although it also provides useful animal fodder in winter. The culms are not very straight and have rather swollen nodes and many branches, but the ease of propagation at subtropical altitudes makes this a very valuable species.

This is a resilient bamboo which can survive on drier sites than species such as *Himalayacalamus hookerianus,* which could provide straighter weaving material in moister sites. It is planted in gullies and on waste land but can also be planted on terrace risers, where it is very effective in soil stabilisation.

The traditional propagation method is very successful in this species, as it produces a large number of new shoots at a fast rate, and the rhizomes can be extracted easily.

Drepanostachyum khasianum   (Dz. *daphe* Nep. *ban nigalo*)   T49

Culm &
culm sheath

- swollen nodes
  with no hairs

- little wax

- culm smooth

Leaf sheath

- long ligule with
  dense hairs

- variable
  deciduous
  auricles

- few hairs

Culm sheath apex

- ligule short

- sheath with no hairs

- interior with a few
  spines or hairs

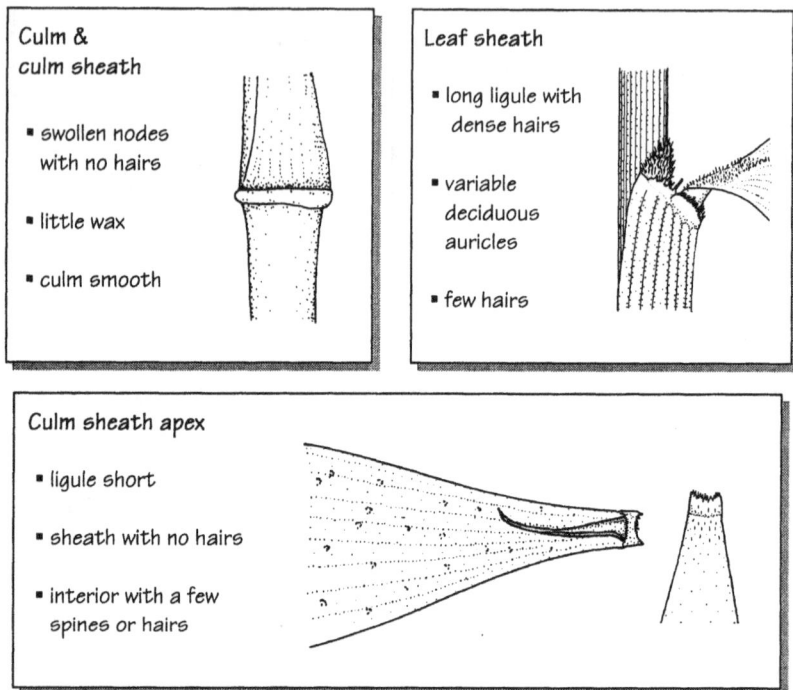

This species has been found in Wangdi and Gasa districts between 1,000m and 1,600m, and it is very common around Thinleygang. It is found in the forest and along roadsides, but it does not seem to be cultivated widely, probably because of the availability of higher quality temperate bamboos in nearby forests, such as *Borinda grossa*.

It is heavily browsed and is usually reduced in size to below 1cm in diameter and 3m in height, although like other *Drepanostachyum* species it could reach 5m in height if not browsed. This species has few prominent characters by which it can be recognised, although the culms are often dark, and the leaves are very small. It is identified by the absence of characters rather than by their presence. The culm sheaths have no hairs, auricles, or bristles, and a short ligule. The leaf sheaths have few hairs, small deciduous auricles, and a ligule which is more pubescent than that of other *Drepanostachyum* species. It can be distinguished from all the species of *Himalayacalamus* by the presence of rough spines inside the culm sheath.

It is identified as *D. khasianum*, a species found in Shillong by Griffith, but this is tentative as its flowers are not yet known.

8. HIMALAYACALAMUS

Clump-forming thornless bamboos up to 8m tall, found from 1,800m to 2,500m in cool broadleaved forest, and also widely cultivated. They have single flowers, short buds, and 15-40 branches. Leaves do not have the cross-veins seen in *Thamnocalamus* and *Borinda* species, and internodes are less than 40cm long, shorter than those of most species of *Cephalostachyum* and *Teinostachyum*. Although similar to species of *Drepanostachyum*, they differ in many ways. Branch buds at mid-culm nodes have fewer initials visible (compare figs. 35 and 31). Branches are fewer, usually around fifteen in the first year. They vary in size, are more erect, and do not spread right around the culm. The basal culm internodes are progressively longer. The culm sheaths are completely smooth inside, and usually broad towards the top, with a short ligule. They are more tolerant of cold, and are often found at higher altitude, but they are also less drought tolerant. The new shoots of several species are edible.

fig. 33 - clump appearance

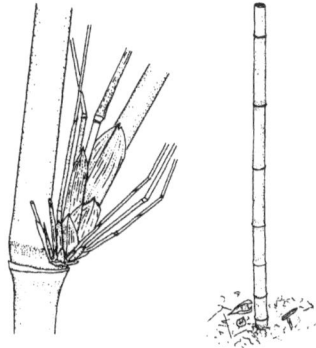

fig. 34 - young branches and culm base

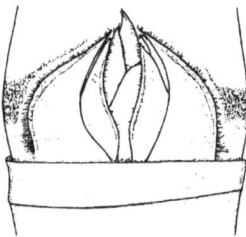

fig 35 - mid-culm branch bud

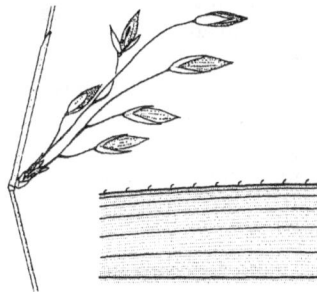

fig 36 - flowers and leaf veins

Himalayacalamus falconeri (Nep. *singhane*)   T27

**Culm &
culm sheath**

- swollen nodes
  with red bands

- broad shoulders

- sheath smooth,
  often striped

Leaf sheath

- rounded ligule
  with hairs

- no auricle

- no hairs
  or bristles

Culm sheath interior

- ligule short and broad

- apex membranous
  with visible cross-veins

- interior smooth, no hairs or spines

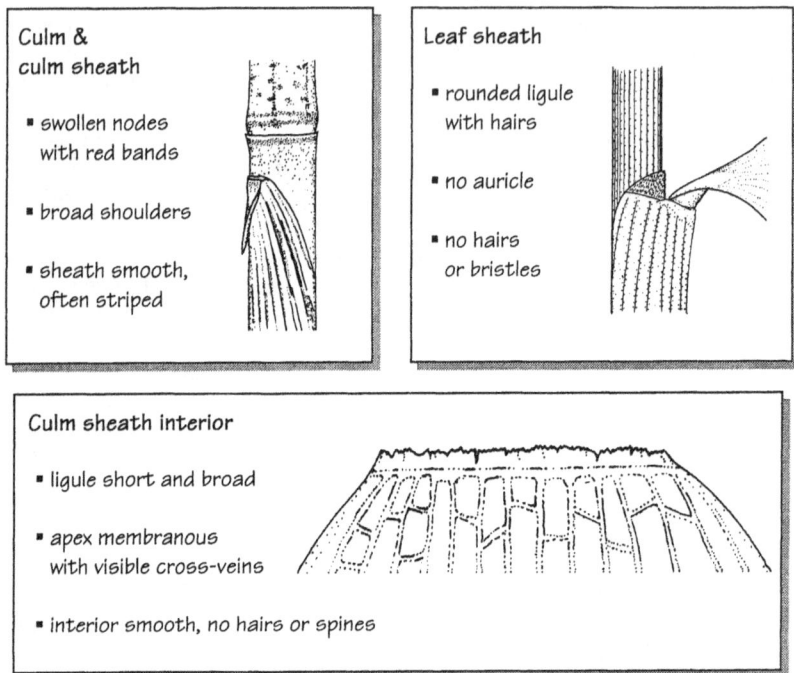

A locally common bamboo in cool broadleaved forests of western Bhutan, between 2,000m and 2,500m.

This species can be distinguished from most others by the absence of hairs or auricles on the sheaths, the smooth culm sheath interior, and the absence of cross-veins on the leaves. The broad culm sheaths are bullet-shaped, with a short broad ligule, and are often striped with yellow and purple lines. The young shoots of this species often have a thick glutinous covering.

H. falconeri usually lacks any distinctive characteristics of its own, but the variety in western Bhutan has slightly ridged culms and light hairs at the base of the culm sheaths.

The shoots of this species are edible and they are collected from the forest, which can sometimes conflict with the use of the older culms for weaving.

Other similar bamboos have been collected in Shemgang and Tashigang districts, but mature culm sheaths were not available, so it is impossible to describe them properly. The Tashigang plants were very interesting as they had roots from the central branch bases. This is very unusual in small bamboos, and it could make propagation by culm cuttings a possibility.

Himalayacalamus hookerianus   (Nep. *padang*)   T4

Culm &
culm sheath

• blue culm with
  light wax

• reflexed blade

• tough sheath,
  narrow apex

Leaf sheath

• margin
  ciliate

• long ligule

• no hairs
  or bristles

Culm sheath apex

• sheath rolled with
  ciliate margins

• long ligule,
  tip not ragged

• interior smooth, no spines or hairs

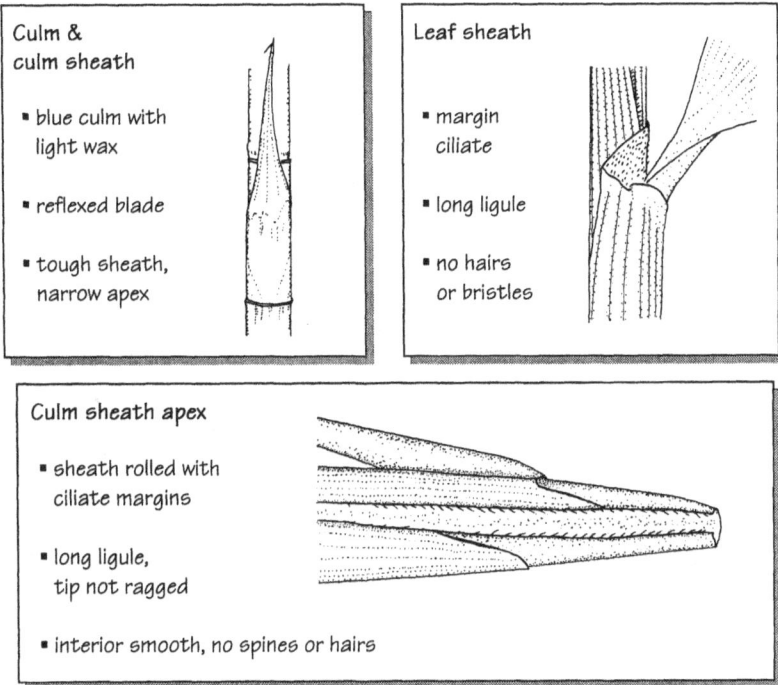

The most common small cultivated bamboo in Chirang district, up to 3cm in diameter and up to 7m tall. It can easily be recognised by the blue colour of the new culms, and by the long narrow necks of the tough culm sheaths. *Drepanostachyum* species have similar culm sheaths but the interior is rough at the top, and they have greener culms. Species from other genera with blue culms have much broader culm sheaths and leaves with cross-veins.

It is found from 1,400m to 2,100m, occurring naturally in the forest in Sikkim, but mainly cultivated in Nepal and Bhutan. The principle use of this species is basket-making, although it also provides useful animal fodder. The culms have fewer branches than those of *Drepanostachyum* species, and they have longer internodes than *Himalayacalamus falconeri*. This makes the culms easier to split into weavable strips.

It is planted in gulleys and on waste land, but can also be planted on terrace risers, where it is very effective in soil stabilisation, although it may grow quite large and shade field crops excessively.

The lack of branches at the base of the culm makes propagation by the traditional technique a little more difficult. A longer pole must be used to ensure the successful development of branches from buds at the top.

9. CHIMONOBAMBUSA

Spreading slightly frost-hardy bamboos with culms up to 6m tall and up to 4cm in diameter, found in cool broadleaved forest types from 1,400m to 2,000m. The culm nodes are raised, with three branches, and often have a ring of prominent thorns. The culms are upright, and branch singly from long rhizomes that have roots at all nodes. They are related to the common spreading bamboos of China and Japan in the genus *Phyllostachys*, but have three branches at mid-culm nodes (fig. 40), instead of two, as well as raised nodes which may bear rings of thorns (fig. 39). The flattening on one side of the culm and the branches is also less developed, and the culm sheath blades are very small. The rhizomes are similar to those of *Arundinaria*, but the culm nodes and branching are very diferent. Species of *Chimonocalamus*, known from China Meghalaya and Sikkim, also have thorns, but they grow in clumps with short rhizomes like those of *Thamnocalamus*, and have not yet been found in Bhutan.

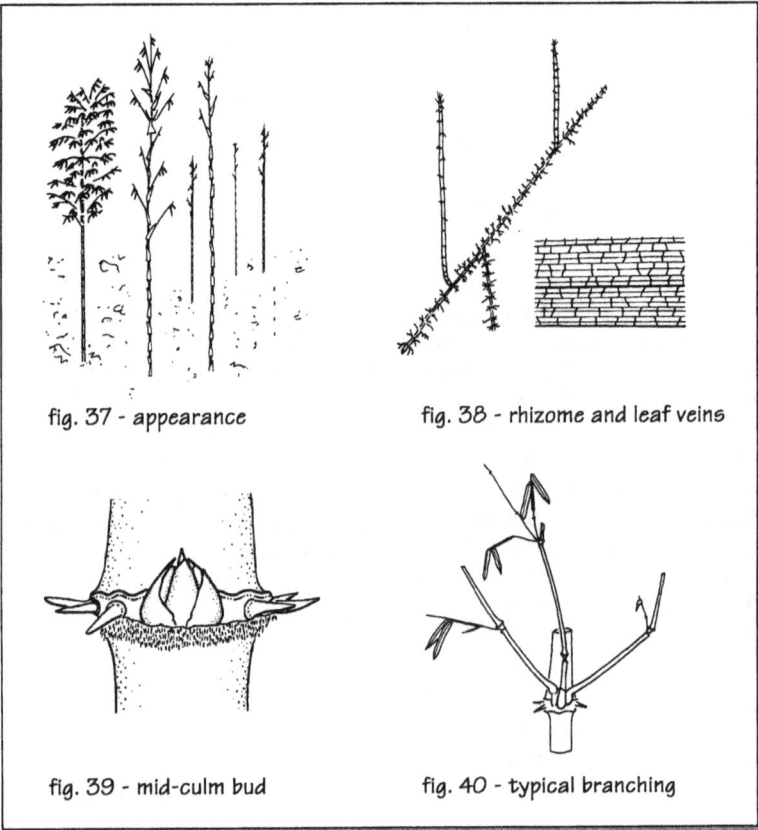

fig. 37 - appearance

fig. 38 - rhizome and leaf veins

fig. 39 - mid-culm bud

fig. 40 - typical branching

Chimonobambusa callosa (Dz. *u*, Nep. *khare bans*) A43

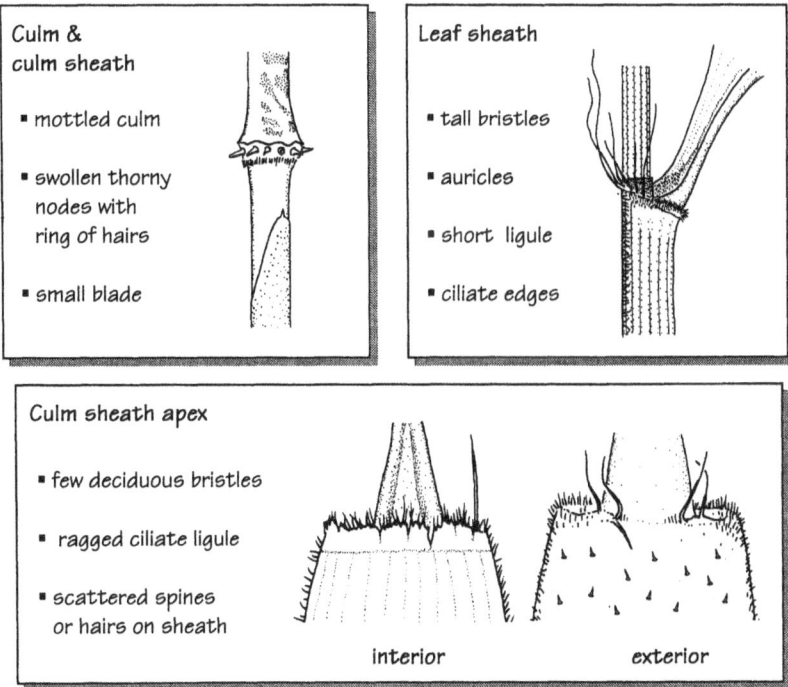

Culm &
culm sheath

- mottled culm

- swollen thorny
  nodes with
  ring of hairs

- small blade

Leaf sheath

- tall bristles

- auricles

- short ligule

- ciliate edges

Culm sheath apex

- few deciduous bristles

- ragged ciliate ligule

- scattered spines
  or hairs on sheath

interior   exterior

A thorny bamboo, easily distinguished from all other Himalayan bamboos by the growth of solitary shoots from long spreading rhizomes, in addition to thorns around the culm nodes.

Even without rhizomes and culms, it can still be recognised by the raised nodes on branches and even on small branchlets, or by the lightly hairy culm sheaths with small blades. Underneath the culm sheaths the culms often have mottled brown stains.

The other bamboo in the region with thorns around the culm, *Chimonocalamus griffithianus* from Sikkim and Meghalaya, may occur in Bhutan, but can be distinguished by having short rhizomes and culms arising in clumps. The Nepali local name *khare bans* is also used for the much larger Indian species *Bambusa arundinacea*, which has thorns developed from shortened branchlets, rather than from aerial roots on the culm nodes.

This species is common throughout the cool wet subtropical broadleaved forest from 1,400m to 2,000m. It is not as dense as *Yushania* spp, having well separated culms and graceful branches of yellow-green foliage.

The culms are strong, but brittle and of little use, with the thorns making them very unpleasant to handle.

10. MELOCANNA

One spreading thornless tropical species introduced from Bangladesh, with straight upright culms up to 21m tall and 7cm in diameter, arising from rhizomes which are up to 2m long. It is only found below 1,400m. This is the largest spreading bamboo in the region, highly valued for straightness, durability, and excellent paper-pulp. The large size of the culms is usually sufficient to distinguish it from other spreading bamboos. The walls are up to 1cm thick, much thicker than the thin walls of *Pseudostachyum*. The culms have no thorns and the nodes are not raised, unlike those of *Chimonobambusa*. The long rhizomes distinguish it from clump-forming bamboos of similar stature. The culm sheath is very distinctive, with a long narrow blade. Culm buds are short and tough, and closed at the front. There are up to 40 branches from each culm node, and they are all similar in size. Leaves have no cross-veins. One species is known, forming extensive stands in Bangladesh.

fig. 41 - appearance

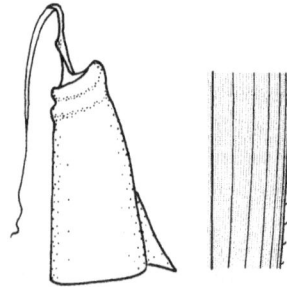

fig. 42 - culm sheath and leaf veins

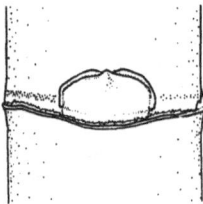

fig. 43 - mid-culm bud

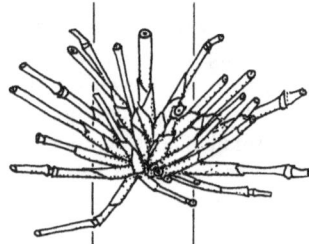

fig. 44 - typical branching

Melocanna baccifera   (Nep. *lahure bans*)

Culm &
culm sheath

• level nodes

• light wax

• long blade

• no hairs

Leaf sheath

• tall bristles

• wide auricles

• short ligule

• narrow petiole

Culm sheath apex

• narrow blade

• serrated ligule
  1-2mm tall

• corrugated sheath

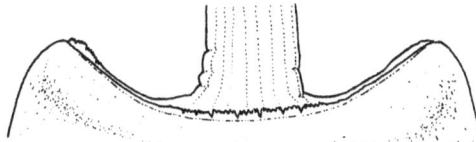

A distinctive bamboo cultivated near the southern border, where it forms graceful open stands of medium-sized very upright culms, reaching 12m in height and 5cm in diameter. It requires high temperatures and rainfall of over 2m per year to reach its maximum potential height of 21m.

The culm sheaths are covered in white hairs at first and have two strong waves towards the top. There is a ridge (callus) on the outside of the sheath where the blade is attached. In other bamboos this is normally only seen on leaf sheaths. The blade is sword-shaped and longer than the sheath. Leaf sheath auricles are prominent with very long erect white wavy bristles. The fruits are famous for their large size and shape, similar to that of a pear, and they may germinate before falling off the mother plant, making storage impossible.

The culms are smaller than those of *Bambusa* or *Dendrocalamus* species, but they are thick-walled (solid at the base), very straight, and said to be termite resistant. They provide a general purpose construction material, and are also widely used for mats.

This species cannot be propagated by culm cuttings. The traditional planting technique is most appropriate, using a short rhizome length - the long rhizome neck is not required.

11. PSEUDOSTACHYUM

A spreading thornless subtropical semi-scandent bamboo, with thin-walled pendulous shoots up to 16m tall, and rhizomes with long solid rootless necks. Distribution is restricted to high rainfall areas, usually from 300m to 1,200m. The rhizomes are quite similar to those of *Melocanna*, with very long rootless necks, but the culms are not always solitary, often arising instead in well-separated groups of a few culms. The branches are also similar to those of *Melocanna*, although fewer in number. The flowers are sometimes very small, similar to those of *Teinostachyum*, but they often develop into larger, densely hairy curving spikelets. The broad deciduous culm sheath blades distinguish it readily from all the other spreading genera. The buds and leaf veins are very similar to those of all the other medium-sized subtropical genera. The thin walls make splitting and weaving very easy. Only one species is known. It is not usually cultivated, as it requires the support of tree branches to grow satisfactorily.

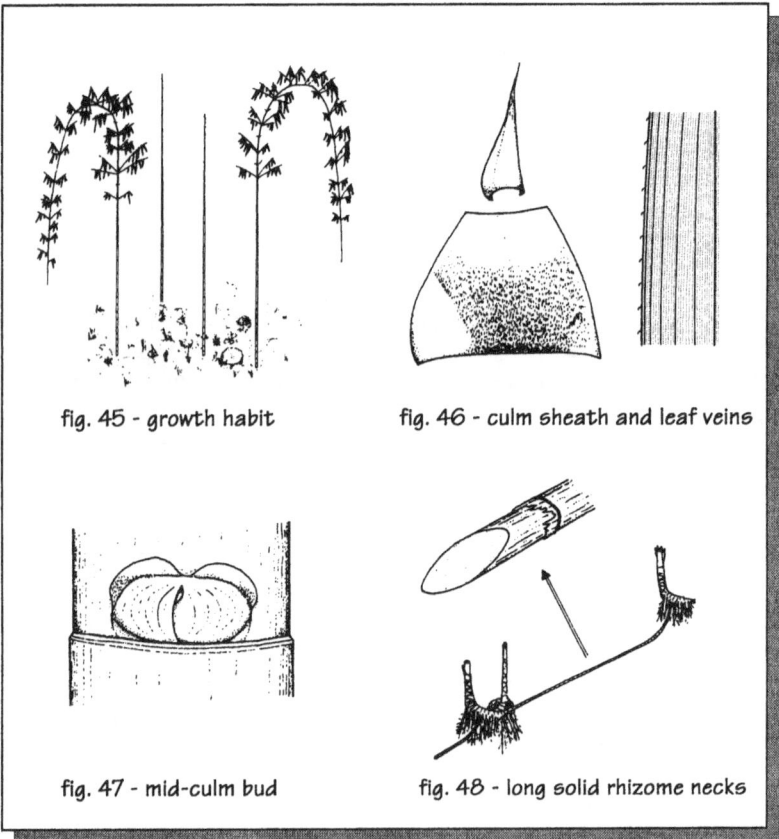

fig. 45 - growth habit

fig. 46 - culm sheath and leaf veins

fig. 47 - mid-culm bud

fig. 48 - long solid rhizome necks

Pseudostachyum polymorphum   (Nep. *philim,* Keng. *dai*)   U44

Culm &
culm sheath

• level unraised
  culm nodes

• no auricle,
  few bristles

• brown hairs

Leaf sheath

• no auricle
  or bristles

• short ligule

• no hairs

Culm sheath apex

• deciduous blade

• upright bristles

• no serrations
  on 1-2mm tall ligule

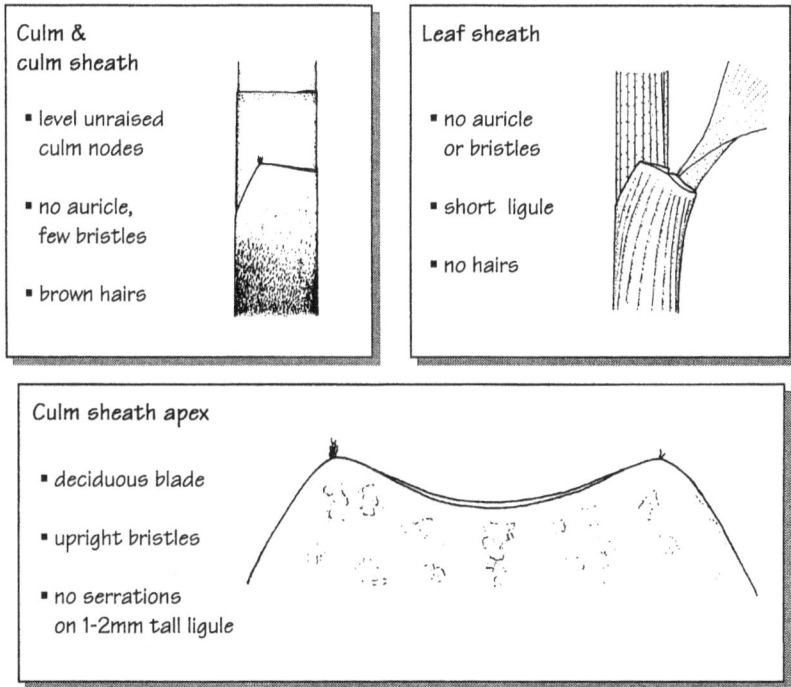

A forest species common in Gayleg-phug and Shemgang districts at around 1,000m with culms scrambling up to 16m in height, but with a diameter of only 4cm or less and culm walls less than 3mm thick.

The delicate culm walls of this species are its most distinctive characteristic. The upright new shoots have dark purple culm sheath blades, which fall off very quickly. Older culms with long pendulous tips can appear very similar to branches of *Dendrocalamus hamiltonii,* and can easily be overlooked. This species spreads widely, but several shoots may arise together at one location, before a long new rhizome extends the clump to a different site. Without tree branches the culms of this species split, and the upper parts of the culms collapse under the weight of their own foliage. The flowers of this species are variable, possibly because of insects.

The main use of this species is weaving, as the thin-walled poles can easily be split or crushed, and woven into panelling, roofing, or fencing. Because of the length of the poles, the absence of branches in the lower parts of the culm, and the unraised nodes, the panelling is strong and can be be woven very tightly. It is reported to be a durable and insect-resistant bamboo, despite the thinness of the culm walls.

12. NEOMICROCALAMUS

Spreading thornless bamboos, often scrambling, from subtropical to warm-temperate areas, with culms less than 2cm in diameter, long shiny internodes, strong central branches, and needle-like culm sheath blades. The flowers appear similar to those of small bamboos such as *Arundinaria*, but they have six stamens instead of three. Therefore they are also related to genera such as *Bambusa*, and may represent an intermediate stage in bamboo evolution. They use trees for support, and in Bhutan are found only in the wettest subtropical forests in the east. The erect needle-like culm sheath blades and the smooth shiny culm surface distinguish these bamboos. The mid-culm buds are tall and narrow. There may be up to eighteen similar branches, or the central branch may be strongly dominant, and similar in size to the culm. The culms are very flexible, and are used for high quality weaving. *Yushania* species can appear similar, but have prominent cross-veins on their leaves and usually have rougher culms.

fig. 49 - appearance

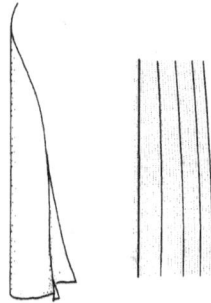

fig. 50 - culm sheath and leaf veins

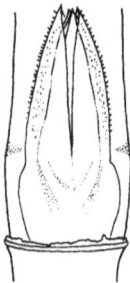

fig. 51 - mid-culm bud

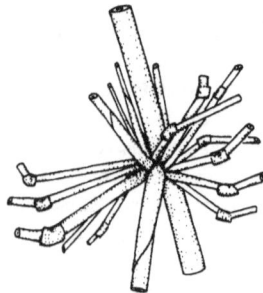

fig. 52 - typical branching

Neomicrocalamus andropogonifolius (Shar. *ringshu*, Keng. *ula*, Nep. *langma*)   S41

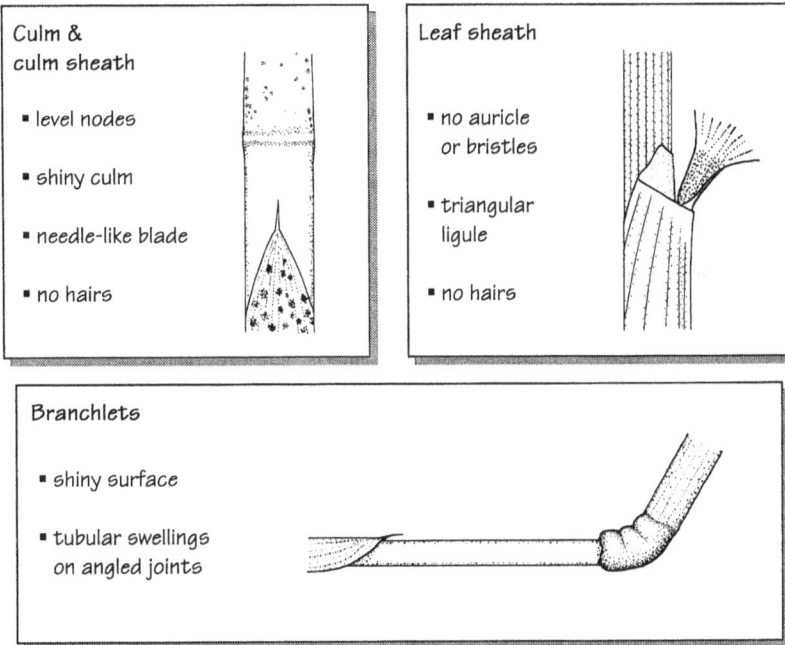

Culm &
culm sheath

• level nodes

• shiny culm

• needle-like blade

• no hairs

Leaf sheath

• no auricle
  or bristles

• triangular
  ligule

• no hairs

Branchlets

• shiny surface

• tubular swellings
  on angled joints

This species is restricted to the wetter eastern forests in Bhutan, especially around 1,600-1,800m, where it is found in conjunction with *Cephalostachyum latifolium* and *Chimonobambusa callosa*. The long scrambling central branches spread over small trees, with sprays of small broad leaves hanging out into the light. The minor branches readjust quickly to changing light direction by growth of small swellings around the joints.

The culms are highly valued because of their flexibility, durability, and hard shiny surface. Split sections are woven into colourful traditional lunch boxes made from two closely fitting bowls (bangchungs), other kinds of baskets, and coverings for containers such as arrow quivers, after dying different strips in bright colours. These containers may be almost watertight as the strips from the surface of the culm are smooth and fit together very closely. The edges of the baskets are sown together using strips of a small cane, which are more flexible than strips of bamboo and can be tied into tight knots.

Other *Neomicrocalamus* species from North-eastern India and Tibet have thicker culm walls or solid culms and rough culm sheaths. This species is also known from Meghalaya. It should be easy to propagate by the traditional technique, but it needs tree branches or bushes for support and shade.

13. YUSHANIA

Spreading thornless frost-hardy bamboos, forming dense thickets or covering large areas, with upright culms from 1m to 4m tall, found in temperate forests and open grazing areas, from 1,800m to 3,600m, often stunted by browsing livestock. Leaves have clear cross-veins, unlike the leaves of the subtropical spreading genera *Melocanna* and *Pseudostachyum,* which only have parallel veins. The culms are not prominently ridged as in *Borinda,* and the branches are fewer in number and more upright The young culms of most species are rough below the nodes, while those of *Thamnocalamus* are always smooth. *Yushania* species have rhizomes of more than 30cm with rootless necks. The rhizomes may be solid or hollow with no dividing walls at their nodes. *Chimonobambusa* and *Arundinaria* species have roots all along the rhizome (figs. 38 and 54), and their hollow rhizomes are closed at each node. When growing strongly, *Yushania* species can prevent the natural regeneration of trees.

fig. 53 - appearance

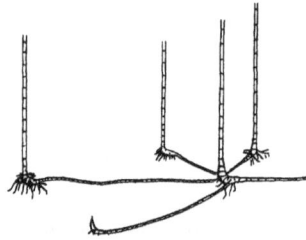

fig. 54 - rhizomes with long necks

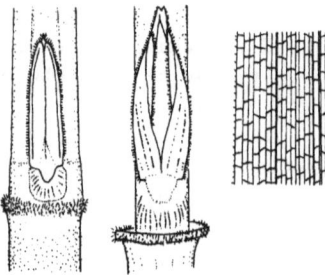

fig. 55 - buds and leaf veins

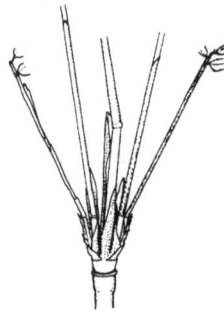

fig. 56 - typical branching

KEY TO *YUSHANIA* SPECIES

Rhizome necks hollow; leaf edges different
with thick clear band along one edge .... *microphylla*

Rhizome necks solid; leaf edges similar, both thin:-

    culm sheath auricles large & persistent
    with many widely spreading bristles .... *hirsuta*

    culm sheath auricles small or absent
    with a few erect bristles:-

        new culm sheath base with dense ring of hairs; leaf
        sheath with short ligule and many long erect bristles .... *pantlingii*

        new culm sheath base with few hairs or none at all;
        leaf sheaths with long ligules and few bristles .... *maling*

Yushania hirsuta   (Dz. *hima*)   T41

Culm &
culm sheath

- tough sheath

- hairy nodes

- rough culm

- large auricles,
  spreading bristles

Leaf sheath

- large auricle

- spreading
  bristles

- long ligule

- long hairs

Rhizome neck

- elongated
- solid internodes
- no roots

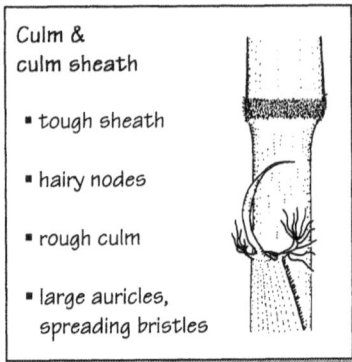

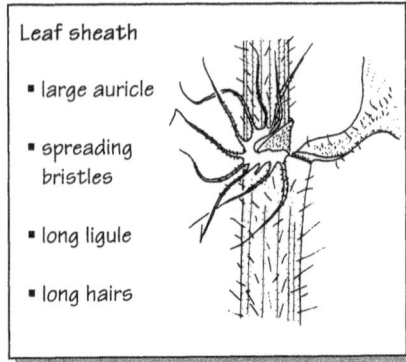

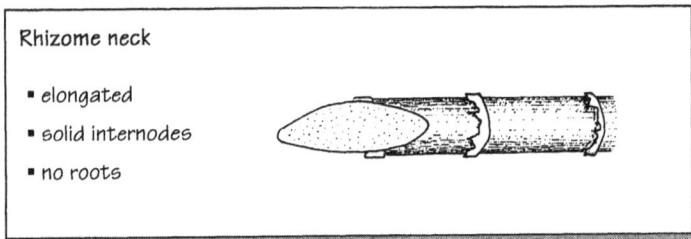

This bamboo is a common component of coniferous and broadleaved forest in western Bhutan especially between 1,800m and 2,800m. It is a large *Yushania* species, reaching up to 8m in height and 2cm in diameter, with internodes up to 40cm long. The combination of a dense ring of hairs at the base of the culm sheaths, very rough culms, and very large spreading antler-shaped auricles distinguish this species from all other *Yushania* species. The lower nodes often have only one branch, which can be up to 1cm in diameter.

It is not usually harvested when other bamboos are available, but it can be used for fencing or eccra walling. It also provides areas of browsing for livestock in the winter.

This species often forms dense dark impenetrable thickets where no other vegetation can survive, and it can severely restrict the regeneration of tree species. The rhizomes are deep, solid and strong, so that physical removal is very difficult. In order to suppress it substantially the shoots must be cut repeatedly each summer. Grazing by livestock and burning may also be effective. Areas of forest with large amounts of this species in particular in the understorey should not really be clear-felled, as the extra light may allow the bamboo to dominate regeneration, and weeding costs will be much higher.

Yushania maling   (Nep. *malingo, maling*)   T9, T53

```
Culm &
culm sheath

• no dense
  ring of hairs

• very rough
  culm

• few bristles
```

```
Leaf sheath

• tall bristles

• auricles small
  or absent

• long ligule
```

```
Rhizome

• neck elongated
• internodes solid
• no roots on neck
```

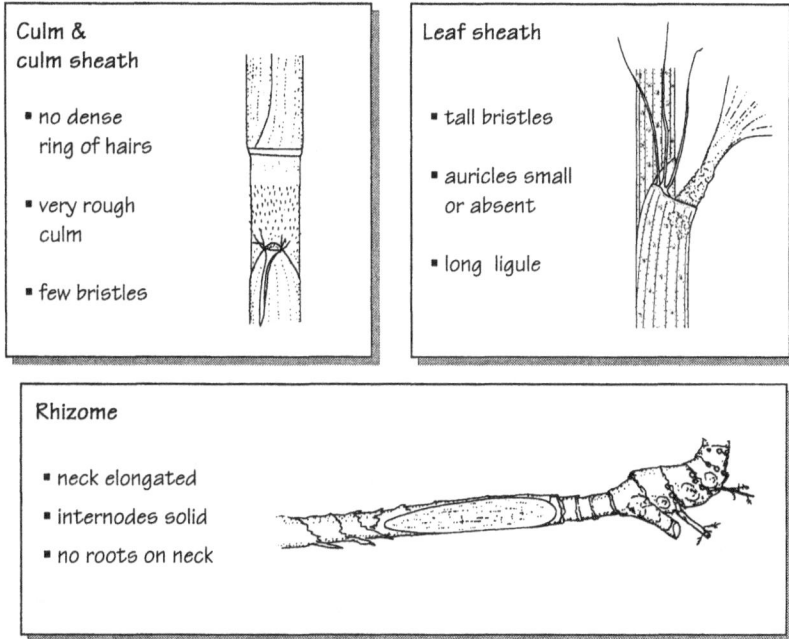

This species is common in East Nepal, Sikkim, and West Bengal and extends into western Bhutan. It occurs from 1,800m to about 3,000m.

It is similar to *Y. pantlingii* of eastern Bhutan, but it seems to be smaller, and does not have a ring of long hairs at the culm sheath base, and has fewer shorter bristles at the top of the leaf sheaths. The solid rhizome necks, and lack of a clear thickened band on one leaf edge distinguish it from *Y. microphylla*. The culms are also much rougher, with long clearly-visible bristles below the nodes of new shoots. In Chhukha district it has more hairs at the base of the culm sheath, finely-grooved culms, and fewer, shorter bristles on the culm sheath and new culms.

When growing vigorously the culms may be used for fencing or eccra walling, but they are usually too small for these uses, although they may be used for brushes, arrows, and straws. It is usually similar in size to *Y. microphylla*, with culms rarely reaching 3m in height or 1cm in diameter, but the stands can sometimes become taller, and dense enough to prevent the regeneration of trees.

This species was confused with *Arundinaria racemosa* for a long time. It can easily be distinguished from that species by the rough internodes of young culms, and by the long rhizome necks with no roots. The Nepali local name may also be used for several other spreading and clump-forming bamboos.

Yushania microphylla   (Dz. *meg, mingma*)   T45

Culm &
culm sheath

* ± ring of hairs

* white/black
  waxy ring

* smooth culm

* ± small auricle

Leaf sheath

* one edge
  thickened

* erect bristles

* truncate ligule

* no hairs

Rhizome neck

* elongated
* hollow internodes
* no roots

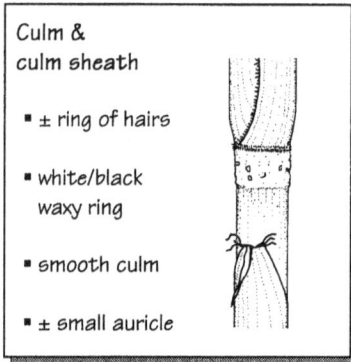

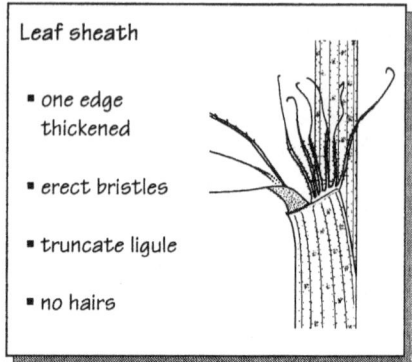

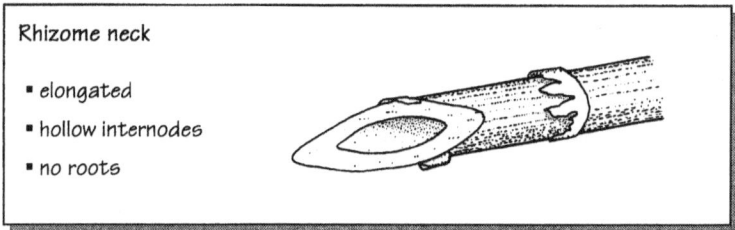

This is a common bamboo of cool temperate areas, extending down to 2,300m in moist gullies. It also reaches the sub-alpine level at 3,500m, and it often forms large areas of grazing land. It is usually found on gently sloping wet areas, rather than steep slopes or gullies.

It is usually heavily browsed and also often burnt, so that it is commonly less than 1m tall, with balls of short branches at each node and leaves of less than 3cm. The culms can be up to 3m tall and 1.5cm in diameter, with leaves of up to 10cm, when protected from grazing.

This species can be distinguished from other *Yushania* species and from *Arundinaria racemosa* by the thick transparent band along one edge of its leaves. It also has a persistent flaky ring of wax below the culm nodes. This changes from white to black with age. In addition, the rhizome necks are hollow, even at the nodes, producing long soft hollow cylinders. Some plants are more hairy than others, and plants from western Bhutan have a ring of hairs at the culm sheath base. New culms can be rough or smooth, and leaf sheath auricles may be absent or pronounced.

This bamboo is often too short to interfere with tree regeneration, and it is important for livestock and wildlife in the winter months, both in open grazing areas, and in the forest. Small shoots may be eaten by flocks of black-necked cranes. The long hollow rhizome necks may assist in drainage and aeration in flat seasonally waterlogged sites.

Yushania pantlingii T56

Culm &
culm sheath

- frill of light
  brown hairs

- slightly
  rough culm

- upright bristles,
  no auricles

Leaf sheath

- thin leaf edges

- tall bristles

- no auricles

- very short
  ligule

Rhizome neck

- elongated
- solid internodes
- no roots

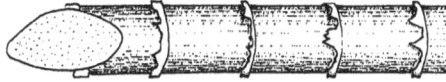

This bamboo is a common component of coniferous and broadleaved forest in temperate parts of central and eastern Bhutan, from 1,700m to 2,600m. It is very similar to the western species *Y. maling,* but it is usually larger.

This is a very vigorous bamboo, reaching up to 8m in height and 2cm in diameter, with internodes up to 40cm long.

The combination of a dense ring of hairs at the base of the culm sheaths, no thickened transparent band on the leaf edges, and erect leaf sheath bristles with no auricles, distinguishes this bamboo from other *Yushania* species in Bhutan.

The culm sheaths and the leaf sheaths are quite similar to those of *Borinda grossa,* but the culms are rough and do not have the fine ridges seen on the culms of *B. grossa,* and it spreads rather than forming tight clumps.

This bamboo can often form dense thickets that severely restrict the regeneration of trees. Like *Y. hirsuta* it can be categorised as an invasive weed. If it is to be suppressed substantially, it has to be burnt and grazed, and the the new shoots should be cut repeatedly each summer. The rhizomes grow deep in the soil, and are solid and strong, so that physical removal is very difficult.

Areas of forest with any large vigorously growing species of *Yushania* in the understorey should not ideally be clear-felled, as the extra light may allow the bamboo to dominate regeneration of tree species.

14. ARUNDINARIA

Spreading thornless frost-hardy small bamboos, with upright culms up to 3m tall, simple branching, and rhizomes with roots at all nodes. They are found in temperate forest and grazing areas, from 2,900 to 3,600m, often mixed with *Yushania* or *Thamnocalamus* species. Similar in appearance to small *Yushania* species, but with culms which are always smooth and have little or no wax. The rhizomes are also very different. They continue under the ground indefinitely, with roots at all the nodes, producing well-separated culms at regular intervals. *Chimonobambusa callosa* has very similar rhizomes, but its culm nodes are raised and may bear thorns, while the nodes of *Arundinaria* species are level. Buds are tall, similar to those of *Yushania* and *Borinda*, but the branching is simpler, with a single branch leaving the culm, then branching repeatedly in a fan-shaped arrangement. The leaves have very prominent cross-veins, unlike those of the two spreading subtropical genera, *Melocanna* and *Pseudostachyum*.

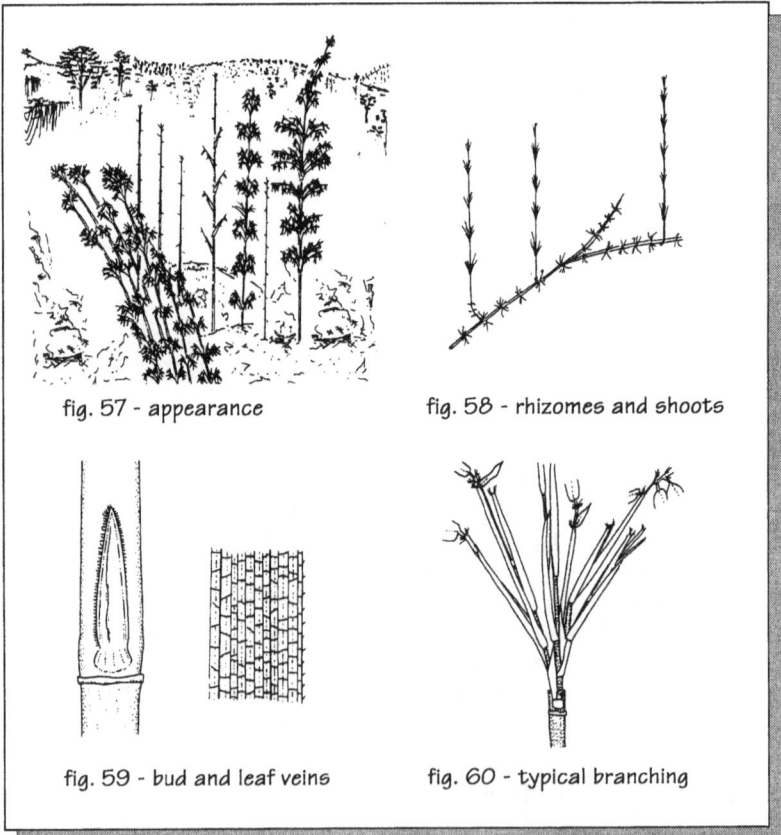

fig. 57 - appearance

fig. 58 - rhizomes and shoots

fig. 59 - bud and leaf veins

fig. 60 - typical branching

Arundinaria racemosa

Culm &
culm sheath

- no hairs

- smooth culm,
  no wax

- spreading bristles,
  small auricles

Leaf sheath

- upright bristles

- tall auricle

- short ligule

- few hairs

Rhizome

- internodes all
  elongated & hollow

- nodes all solid,
  & all bearing roots

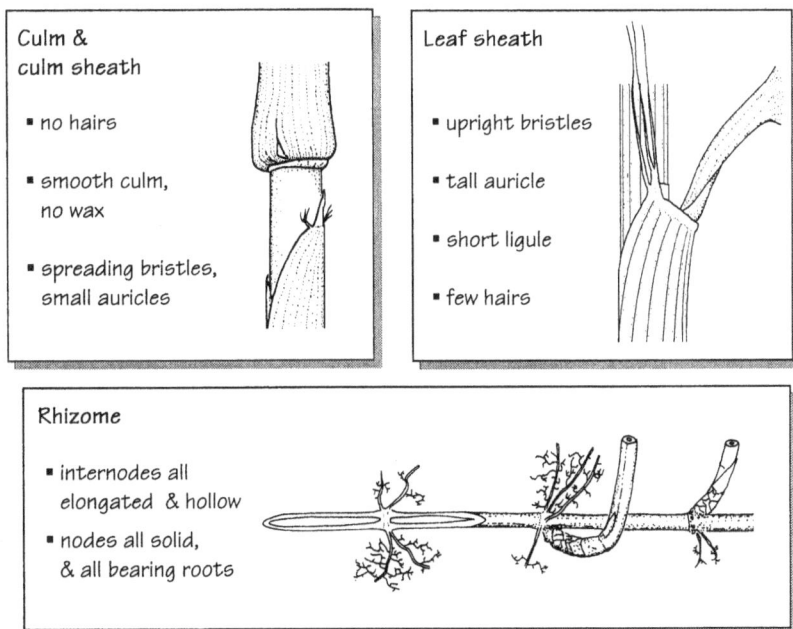

This is a common bamboo of the higher altitude coniferous forests above 2,900m, often forming a component of mixed bamboo pastureland. It is usually found in better drained or more sloping sites than *Yushania* species, but not usually on the steepest sites where *Thamnocalamus spathiflorus* thrives.

It is usually less than 2m tall and 1cm in diameter, with leaves up to 10cm long, but like *Yushania* species it is often stunted by grazing and burning. Larger plants may be found, but it does not usually form dense thickets.

This species is best distinguished from *Yushania* species by the totally different form of rhizome, with roots growing at every node. It can be distinguished quickly from *Yushania microphylla*, another common spreading species with which it is often found, by the absence of any thick transparent band along the edges of its leaves. It also has fewer, stronger, less scabrous bristles on the leaf sheath auricles. New culms are always smooth, without any roughness or wax below the nodes.

This species is usually too small to shade out tree regeneration, and tends to form more open stands than *Yushania* species. It is important for grazing of livestock, and for wildlife, and may also be used for making arrows, brushes and drinking straws.

CHECKLIST OF SPECIES AND AUTHORITIES
(with synonyms in italics)

Ampelocalamus Chen, Wen & Sheng
    A. patellaris (Gamble) Stapleton
        Dendrocalamus patellaris Gamble
        Chimonobambusa jainiana Das & Pal
        Drepanostachyum jainianum (Das & Pal) R.B. Majumdar

Arundinaria Michaux
    A. racemosa Munro
        Fargesia racemosa (Munro) Yi
        Yushania racemosa (Munro) R.B. Majumdar

Bambusa Schreber
    B. alamii Stapleton
    B. balcooa Roxburgh
        Dendrocalamus balcooa (Roxburgh)Voigt
    B. clavata Stapleton
    B. nutans Wallich ex Munro **subsp. cupulata** Stapleton
        Bambusa macala Wallich
    B. tulda Roxburgh
        Dendrocalamus tulda (Roxburgh) Voigt
    B. vulgaris Schrader ex Wendland

Borinda Stapleton
    B. grossa (Yi) Stapleton
        Fargesia grossa Yi

Cephalostachyum Munro
    C. latifolium Munro
        Schizostachyum latifolium (Munro) R.B. Majumdar
        Cephalostachyum fuchsianum Gamble
        Schizostachyum fuchsianum (Gamble) R.B. Majumdar

Chimonobambusa Makino
    C. callosa (Munro) Nakai
        Arundinaria callosa Munro
        Chimonocalamus callosus (Munro) Hsueh & Yi

Dendrocalamus Nees

>**D. giganteus** Munro
>
>**D. hamiltonii** Munro **var hamiltonii**
>
>**D. hamiltonii** Munro **var. edulis** Munro
>
>**D. hookeri** Munro
>
>**D. sikkimensis** Gamble ex Oliver

Drepanostachyum Keng f.

>**D. annulatum** Stapleton
>
>**D. intermedium** (Munro) Keng f.
>
>>*Arundinaria intermedia* Munro
>>
>>*Chimonobambusa intermedia* (Munro) Nakai
>>
>>*Sinarundinaria intermedia* (Munro) Chao & Renvoize
>
>**D. khasianum** (Nees) Keng
>
>>*Arundinaria khasiana* Munro
>>
>>*Chimonobambusa khasiana* (Munro) Nakai

Himalayacalamus Keng f.

>**H. falconeri** (Munro) Keng f.
>
>>*Thamnocalamus falconeri* Munro
>>
>>*Arundinaria falconeri* (Munro) Benth. & Hook. f
>>
>>*Drepanostachyum falconeri* (Munro) McClintock
>>
>>*Fargesia collaris* Yi
>>
>>*Fargesia gyirongensis* Yi
>
>**H. hookerianus** (Munro) Stapleton
>
>>*Arundinaria hookeriana* Munro
>>
>>*Sinarundinaria hookeriana* (Munro) Chao & Renvoize
>>
>>*Chimonobambusa hookeriana* (Munro) Nakai
>>
>>*Drepanostachyum hookerianum* (Munro) Keng f.

Melocanna Trinius

>**M. baccifera** (Roxburgh) Kurz
>
>>*Melocanna bambusoides* Trin.

Neomicrocalamus Keng f.

>**N. andropogonifolius** (Griffith) Stapleton
>
>>*Bambusa andropogonifolia* Griffith

59

Pseudostachyum Munro

 P. polymorphum Munro

  Schizostachyum polymorphum (Gamble) R.B. Majumdar

Teinostachyum Munro

 T. dullooa Gamble

  Neohouzeaua dullooa (Gamble) Camus

  Schizostachyum dullooa (Gamble) R.B. Majumdar

Thamnocalamus Munro

 T. spathiflorus (Trin.) Munro **subsp. spathiflorus**

  Arundinaria spathiflora Trinius

  Arundinaria aristata Gamble

  Thamnocalamus aristatus (Gamble) E.G. Camus

  T. spathiflorus subsp. *aristatus* (Gamble) McClintock

 T. spathiflorus (Trin.) Munro **var. bhutanensis** Stapleton

Yushania Keng f.

 Y. hirsuta (Munro) R.B. Majumdar

  Arundinaria hirsuta Munro

  Sinarundinaria hirsuta (Munro) Chao & Renvoize

 Y. maling (Gamble) R.B. Majumdar

  Arundinaria maling Gamble

  Sinarundinaria maling (Gamble) Chao & Renvoize

 Y. microphylla (Munro) R.B. Majumdar

  Arundinaria microphylla Munro

  Sinarundinaria microphylla (Munro) Chao & Renvoize

 Y. pantlingii (Gamble) R.B. Majumdar

  Arundinaria pantlingii Gamble

  Semiarundinaria pantlingii (Gamble) Nakai,

  Butania pantlingii (Gamble) Keng f.

  Sinarundinaria pantlingii (Gamble) Chao & Renvoize

GLOSSARY

Glossary of technical terms

aerial root	a root growing above the ground, in the air
auricle	an ear-like projection at the top of a sheath, fig. 1
blade	leaf, or the equivalent section at the top of a culm sheath, fig. 1
callus	small flaps at top of leaf sheath below petiole
chevron	pattern of V-shaped stripes
cilia	hairs along an edge
ciliate	with hairs along the edge
clump	a collection of many culms growing close together
cross-veins	short veins running across the leaf seen when looking through a leaf held up to the light
culm	the stem or stalk of a grass plant, a pole in large bamboos
dbh	culm diameter measured 1.3m above the ground (breast height)
genus	a group of similar species with the same generic name e.g. *Bambusa*
initials	small parts of a bud which will grow into separate branches
internode	the section of a culm between two nodes
ligule	a projecting tongue where sheath and blade meet, fig. 1
long veins	veins running along the length of the leaf
node	ring around the culm joints where the sheath is attached
petiole	narrow neck between leaf blade and leaf sheath
pulvinus	swelling at base of petiole turning blade to the light
reflexed	bent backwards at more than 90°
rhizome	horizontal underground stem producing roots and new shoots
serrated	like the edge of a saw
scabrous	surface rough to touch with small sharp points
species	a group of similar plants called by the same species name e.g. *strictus*
spreading	not growing in clumps
subspecies	division of a species covering a large geographical area
truncate	straight as though cut off
variety	division of a species found in a small geographical area

Language abbreviations

Dz.	Dzongkha
Nep.	Nepali
Shar.	Sharchop
Keng.	Kengkha

BIBLIOGRAPHY

CHAO, C. S. & RENVOIZE, S. A. (1989). A revision of the species described under Arundinaria (Gramineae) in Southeast Asia and Africa. *Kew Bull.* 44(2): 349–367.

GAMBLE, J. S. (1896). The Bambuseae of British India. *Ann. Roy. Bot. Gard. (Calcutta)* 7(1):1–133

JACKSON, J.K. (1987). *Manual of Afforestation in Nepal.* Forestry Research Project Kathmandu.

KENG, P. C. (1982–3). A revision of genera of bamboos from the world. *J. Bamboo Res.* 1(1):1–19; 1(2):31–46; 2(1):11–27; 2(2):1–17.

MAJUMDER, R. B. (1989). In Karthikeyan et al., *Flora Indicae, Enumeratio Monocotyledonae.* 274-283. Botanical Survey of India, Howrah, Calcutta.

McCLURE, F. A. (1966). *The bamboos: a fresh perspective.* Harvard University Press, Cambridge, Mass.

___ (1973). Genera of bamboos native to the new world. *Smithsonian Contr. Bot.* 9: 1–148.

MUNRO, W. (1868). A monograph of the Bambusaceae. *Trans. Linn. Soc. London* 26:1–157.

NAPIER, I. & ROBBINS, M. (1989). *Forest seed and nursery practice in Nepal.* Forestry Research Project, Kathmandu.

SODERSTROM, T. R. & ELLIS, R. P. (1987). The position of bamboo genera and allies in a system of grass classification. In Soderstrom et al.(eds.). *Grass Systematics and Evolution:* 225-238. Smithsonian Institution Press.

STAPLETON, C. M. A. (1991). A morphological investigation of some Himalayan bamboos with an enumeration of taxa in Nepal and Bhutan. Unpublished PhD thesis, University of Aberdeen.

STAPLETON, C. M. A. (1994a). The bamboos of Nepal and Bhutan Part I: *Bambusa, Dendrocalamus, Melocanna, Cephalostachyum, Teinostachyum,* and *Pseudostachyum* (Gramineae: Poaceae, Bambusoideae). *Edinb. J. Bot.* 51(1):1-32

STAPLETON, C. M. A. (1994b). The bamboos of Nepal and Bhutan Part II: *Arundinaria, Thamnocalamus , Borinda,* and *Yushania* (Gramineae: Poaceae, Bambusoideae). *Ed. J. Bot.* 51(2):

STAPLETON, C. M. A. (1994c). The bamboos of Nepal and Bhutan Part III: *Drepanostachyum, Himalayacalamus, Ampelocalamus, Neomicrocalamus,* and *Chimonobambusa* (Gramineae: Poaceae, Bambusoideae). *Ed. J. Bot.* 51(3):

INDEX

www.ingramcontent.com/pod-product-compliance
Lightning Source LLC
Chambersburg PA
CBHW032016190326
41520CB00007B/504